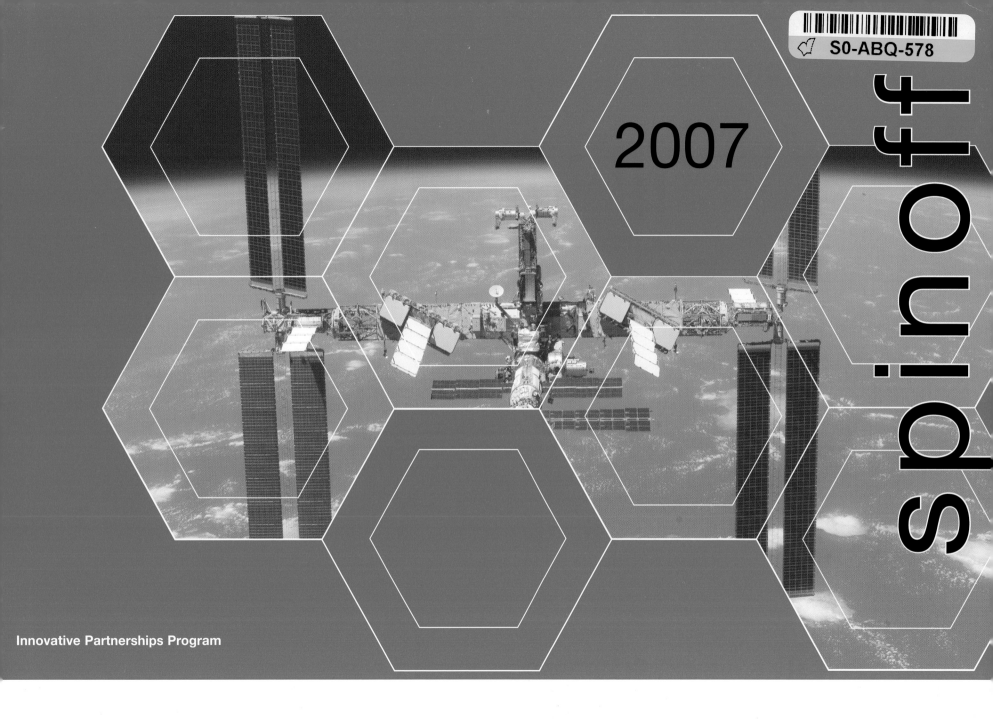

S0-ABQ-578

2007

spinoff

Innovative Partnerships Program

Developed by
Publications and Graphics Department
NASA Center for AeroSpace Information (CASI)

For sale by the Superintendent of Documents, U.S. Government Printing Office
Internet: bookstore.gpo.gov; Phone: toll free (866) 512-1800; DC area (202) 512-1800
Fax: (202) 512-2250 Mail Stop: IDCC, Washington, DC 20402-0001

ISBN 978-0-16-079740-8

On the Cover: Photograph of the International Space Station taken from the Space Shuttle Atlantis on June 19 2007, at the end of STS-117. The mission delivered the second and third starboard truss segments (S3/S4) and another pair of solar arrays to the station.

Table of Contents

Environmental and Agricultural Resources

Computer Technology

Industrial Productivity

Foreword

In our increasingly competitive global economy, strategic U.S. leadership in aeronautics research and space exploration is a critical component of America's strength and vitality. Through our focused work, the men and women of NASA are dedicated to expanding civilization's exploration and scientific horizons in the air and in space and to bringing the inner solar system into our sphere of commerce. I am proud to lead an agency that draws on the exploration imperative found in the human soul to stoke the fires of inspiration and innovation in our citizenry.

The vibrancy of NASA's work can be seen in our current activities, ongoing plans for the future, and international recognition of our achievements.

In 2007, during the space shuttle's 25th anniversary year, three shuttle missions advanced construction work on the International Space Station, adding significantly to the capabilities of this orbiting research outpost and setting the stage for continued expansion of the station's size and research capabilities. We also announced plans for a fifth shuttle servicing mission to the Hubble Space Telescope, now slated for 2008, to extend and improve the telescope's capabilities through 2013.

Looking forward, NASA took major steps to realize the national goal of establishing a permanent base on the lunar surface by 2024 in cooperation with many space-faring nations. In August 2006, we selected a prime contractor to build the Orion crew exploration vehicle, America's next human-rated spacecraft, to be operational by 2015. Last fall, our next-generation launch vehicle, the Ares I, successfully completed its systems requirement review. In December, NASA unveiled an initial Global Exploration Strategy and lunar architecture which details how sustained lunar exploration will advance science, commerce, and technology development and help us prepare for later journeys to Mars and other destinations.

Throughout the year, NASA orbiting spacecraft and rovers continued to expand our understanding of the Red Planet and lay the groundwork for future surface missions.

Also, NASA launched the New Horizons spacecraft to Pluto and twin STEREO spacecraft that produce three-dimensional views of the Sun; recovered comet and interstellar dust particles from the successful Stardust mission; and through our Cassini mission, possibly discovered evidence of liquid water reservoirs that erupt in Yellowstone-like geysers on Saturn's moon Enceladus.

Closer to home, NASA's Earth science team launched the CloudSat and CALIPSO spacecraft to study the role that clouds and aerosols play in regulating Earth's weather, climate, and air quality. And we restructured our aeronautics research portfolio to return to long-term, cutting-edge, fundamental research required to enable the next generation air transportation system and to support our future space missions.

Finally, the entire NASA community cheered the awarding in December of the 2006 Nobel Prize in Physics to Dr. John C. Mather, senior astrophysicist and senior project scientist at the Goddard Space Flight Center. Mather, the first NASA civil-servant employee to win the Nobel Prize, was honored along with George Smoot of the University of California at Berkeley for the "discovery of the black body form and anisotropy of the cosmic microwave background radiation." Mather coordinated the science work of NASA's Cosmic Background Explorer satellite, launched in 1989, which helped validate the Big Bang theory of the origin of the universe.

Fundamental scientific accomplishments like those of Dr. Mather and his colleagues remind us that there is no absolute requirement that NASA produce tangible, practical benefits for the public in everything we do. It is a happy byproduct of our work, however, that many of NASA's missions and activities power innovation that creates new jobs, new markets, and new technologies. Consistent with our Agency's charter, *Spinoff* 2007 highlights NASA's work to "research, develop, verify, and transfer advanced aeronautics, space, and related technologies." Among the useful NASA-derived technologies

Michael D. Griffin
Administrator
National Aeronautics and
Space Administration

featured in *Spinoff* 2007 already receiving prominent use in the commercial and public sectors are:

- A revolutionary method that makes the manufacture of carbon nanotubes safer and less expensive for researchers now creating next-generation electronics.

- NASA-developed air traffic management software tools that are helping to streamline the flow of commercial flights across the entire National Airspace System.

- A new, commercial, all-natural nutritional fat replacement and flavor enhancement product designed with help from NASA's astronaut nutrition program that is now making everyday foods healthier.

I often compare our Nation's investment in space as being equivalent to an individual using a small proportion of their retirement account for a stock that contains the prospect of high risk and high reward. Great nations such as the United States can afford to keep their eyes on the heavens and invest in their long-term future, even as they address the public's more day-to-day concerns. From such vision emanates the hope of continued leaps of progress that will significantly enhance our material lives and lift our collective spirits.

Spinoff (spin´ôf´) -noun.

1. A commercialized product incorporating NASA technology or "know how" which benefits the public. Qualifying technologies include:

   - Products or processes designed for NASA use, to NASA specifications, and then commercialized.

   - Components or processes involving NASA technology incorporated into a commercial product, employed in the manufacturing of a product, or used to modify the design of an existing product.

   - Products or processes to which NASA laboratory personnel made significant contributions, including the use of NASA facilities for testing purposes.

   - Successful entrepreneurial endeavors by ex-NASA employees whose technical expertise was developed while employed by NASA.

   - Products or processes commercialized as the result of a NASA patent license or waiver.

   - Commercial products or processes developed as a result of the Small Business Innovation Research or Small Business Technology Transfer programs.

2. NASA's premier annual publication, featuring successfully commercialized NASA technologies.

Introduction

As NASA approaches its 50th anniversary, we reflect on a proud history of achievements that have pushed back boundaries and opened new frontiers for all humanity in the exploration of our solar system and our understanding of the universe and our place in it. New technologies have arisen during this journey, developed of necessity and spurred by the innovative spirit that answers the call of doing the impossible. In the wake of a half-century of advancement, myriad technologies—without which the limits of aeronautics and space could not have been challenged—have found other uses. From the mundane to the sublime, these technologies have become part of the fabric of our everyday life, driving innovation, helping the economy, and adding to the quality of life not only in the United States, but around the world. This 31st edition of *Spinoff* continues the annual tradition of documenting a few of these happy byproducts by showcasing numerous technologies that have been derived from such diverse projects as launching space shuttles, developing the next generation of helicopters, and analyzing images from Mars probes, all of which are now put to use in our hospitals, homes, and communities.

NASA's core conviction has long been that space exploration is a lens that serves to sharply focus the development of key technologies in a way that simply would not occur without the rigorous scientific demands that arise from seeking to accomplish the near-impossible. Fundamentally, we help to create new technologies to meet our challenging aeronautics and Space Program goals, and once proven, these technologies often have been demonstrated to have a multitude of productive uses in society. Making these sometimes surprising connections and transferring NASA's technologies to the public is an important focus of NASA's Innovative Partnerships Program (IPP).

IPP also seeks solutions to some of NASA's pressing technical challenges by funding Small Business Innovation Research and Small Business Technology Transfer projects, seeking cost-shared, technology-development partnerships through the IPP Seed Fund, and tapping into new sources of innovation through NASA's Centennial Challenges prize program. In seeking innovative partnerships, IPP is moving technology in two directions—both into and out of NASA. IPP achieves these objectives through a nationwide network of offices across NASA's 10 field centers.

As a direct result of the National Aeronautics and Space Act of 1958 and the ensuing Technology Utilization Act of 1962, NASA began to publish *Spinoff*. Since 1976, this widely read and highly anticipated annual has featured life-saving advances in health care, cutting-edge safety innovations for the aeronautics industry, cost-saving manufacturing methods, groundbreaking ecological advances, and a multitude of consumer goods that changes the way we live. Each issue includes dozens of articles highlighting how these products emerged from NASA's investment in aeronautics and space technology to how they are helping real people in everyday life, as well as helping to maintain the position of the United States as the world leader in advanced technologies. *Spinoff* 2007 features such achievements as:

- A network of environmental sensors based on devices worn by astronauts in space and now providing health workers in urban centers with up-to-date information about water quality and disease outbreak information from remote, hard-to-reach areas

- A simple and affordable system for doctors to use ultrasound to perform advanced, noninvasive heart monitoring based on software designed to interpret spacecraft imagery

- An ecologically friendly oil additive originally designed for the space shuttle crawlers at Kennedy Space Center, a wildlife refuge

Spinoff also features such award-winning technologies as:

- The Macro-Fiber Composite, named "NASA Invention of the Year" for 2007, is an innovative, low-cost

Douglas A. Comstock
Director
Innovative Partnerships Program

device designed to control vibration, noise, and deflections in composite structural beams and panels, now used in over 120 different products, including sporting goods, automobiles, aircraft, and spacecraft.

- A portable hyperspectral device designed to analyze differences in radiated energy and monitor temperature and climate change and now used worldwide in the medical, food safety, forensics, counter-terrorism, and military markets has been inducted into the Space Technology Hall of Fame.

- Two additional NASA technologies previously included in *Spinoff* were also inducted into the Space Technology Hall of Fame this year. Emulsified Zero-Valent Iron (*Spinoff* 2005) cleans up contaminated groundwater, and the Microbial Check Valve (*Spinoff* 2006) provides clean drinking water in locations around the globe.

As we look forward to NASA's next 50 years, boundaries will be pushed even further. Future achievements will be enabled by new technologies—many perhaps not even imagined today. Fifty years ago, we could not foresee that heart disease would one day be detected early and with no discomfort, using technology developed to analyze images from Mars probes. So today, we cannot predict the technologies of tomorrow or the spinoffs they will leave in their wake. But as the examples in *Spinoff* 2007 attest, we know they will be there, and we will be looking for them.

Executive Summary

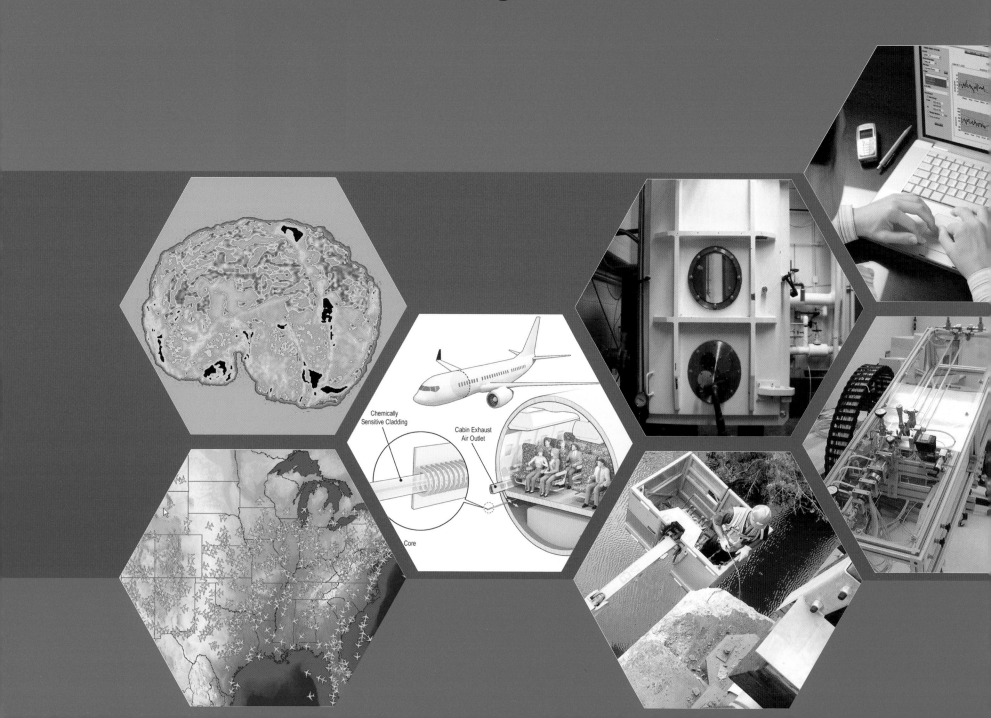

In accordance with congressional mandates cited in the National Aeronautics and Space Act of 1958 and the Technology Utilization Act of 1962, NASA was directed to encourage greater use of the Agency's knowledge by providing a link between the NASA research community and those who might use the research for commercial or industrial products. For more than 40 years, NASA has nurtured partnerships with the private sector to facilitate the transfer of NASA-developed technologies. The benefits of these partnerships have reached throughout the economy and around the globe, as the resulting commercial products contributed to the development of services and technologies in the fields of health and medicine, transportation, public safety, consumer goods, environmental resources, computer technology, and industry. Since 1976, NASA *Spinoff* has profiled more than 1,500 of the most compelling of these technologies, annually highlighting the best and brightest of partnerships and innovations. Building on this dynamic history, NASA partnerships with the private sector continue to seek avenues by which technological achievements and innovations gleaned among the stars can be brought down to benefit our lives on Earth.

Executive Summary

N ASA *Spinoff* highlights the Agency's most significant research and development activities and the successful transfer of NASA technology, showcasing the cutting-edge research being done by the Nation's top technologists and the practical benefits that come back down to Earth in the form of tangible products that make our lives better. The benefits featured in this year's issue include:

Health and Medicine

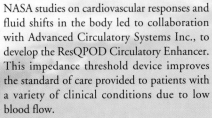

Circulation-Enhancing Device Improves CPR

NASA studies on cardiovascular responses and fluid shifts in the body led to collaboration with Advanced Circulatory Systems Inc., to develop the ResQPOD Circulatory Enhancer. This impedance threshold device improves the standard of care provided to patients with a variety of clinical conditions due to low blood flow.
page 22

Noninvasive Test Detects Cardiovascular Disease

NASA-developed Video Imaging Communication and Retrieval software laid the groundwork for a project seeking to use imaging technology to analyze X-ray images of soft tissue. The same methodology applied to ultrasound imagery resulted in a noninvasive diagnostic system with the ability to accurately predict heart health.
page 26

Scheduling Accessory Assists Patients with Cognitive Disorders

NASA research and funding of Recom Technologies Inc. on artificially intelligent planning reaction models led to the development of a tool to help individuals suffering from various forms and levels of brain impairment. Attention Control Systems Inc. was founded to market the finished device, called the Planning and Execution Assistant and Trainer.
page 30

Neurospinal Screening Evaluates Nerve Function

NASA-funded research in surface electromyography to measure the muscle activity of astronauts was continued by the Chiropractic Leadership Alliance to identify additional applications, resulting in the Insight Subluxation Station, a neurospinal screening and evaluation system. The system, which can be used on infants, children, and adults, is readily compatible with most chiropractic offices and cleared by the U.S. Food and Drug Administration (FDA).
page 32

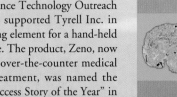

Hand-Held Instrument Fights Acne, Tops Over-the-Counter Market

NASA's Space Alliance Technology Outreach Program (SATOP) supported Tyrell Inc. in redesigning a heating element for a hand-held acne-fighting device. The product, Zeno, now the highest selling over-the-counter medical device for acne treatment, was named the "SATOP Texas, Success Story of the Year" in 2006, Allure's 2005 "Best of Beauty," Marie Claire's "10 Best Gadgets for Girls," and Popular Science's 2005 "Best of What's New." A variation for use in treating herpetic lesions is currently undergoing FDA trials.
page 34

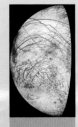

Multispectral Imaging Broadens Cellular Analysis

A NASA Small Business Innovation Research (SBIR) program helped Amnis Corporation develop its ImageStream system to produce sensitive fluorescence images of cells in flow, allowing users to view whole cells rather than just one section of a cell. ImageStream is also built for other applications, including cell signaling and pathway analysis; classification and characterization of peripheral blood mononuclear cell populations; quantitative morphology; apoptosis (cell death) assays; gene expression analysis; analysis of cell conjugates; molecular distribution; and receptor mapping and distribution.
page 36

Hierarchical Segmentation Enhances Diagnostic Imaging

NASA's Recursive Hierachical Segmenting (RHSEG) software, two pattern-matching programs, and three data-mining and edge-detection programs were licensed by Bartron Medical Imaging LLC (BMI) to create the Med-Seg imaging system to analyze CAT and PET scans, MRI, ultrasound, digitized X-rays, digitized mammographies, dental X-rays, soft tissue analyses, moving object analyses, and soft-tissue slides. BMI and NASA are developing a 3-D version of RHSEG to produce pixel-level views of a tumor or lesion, and identify plaque build-up in arteries and density levels of microcalcification in mammographies.
page 40

Transportation

Comprehensive Software Eases Air Traffic Management

NASA's Future Air Traffic Management Concepts Evaluation Tool, developed for air traffic control centers to improve the safety and efficiency of the National Airspace System software, was licensed by Flight Explorer Inc. The software now offers automatic alerts of events such as weather conditions and potential airport delays, and real-time flight coverage over Canada, the United Kingdom, New Zealand, and sections of the Atlantic and Pacific Oceans. Flight Explorer Inc. recently formed several partnerships to expand coverage worldwide.
page 44

Modeling Tool Advances Rotorcraft Design

SBIR contracts helped Continuum Dynamics Inc. develop the Comprehensive Hierarchical Aeromechanics Rotorcraft Model (CHARM) tool for studying helicopter and tiltrotor unsteady free wake modeling, including distributed and integrated loads and performance prediction. CHARM has been used to model a broad spectrum of rotorcraft attributes, including performance, blade loading, blade-vortex interaction noise, air flow fields, and hub loads, and is currently in use by all major rotorcraft manufacturers, NASA, the U.S. Army, and the U.S. Navy.
page 46

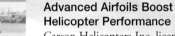

Air Data Report Improves Flight Safety

Sagem Avionics Inc. licensed two NASA-developed software systems: Morning Report, designed to detect atypicalities without pre-defined parameters, and the Aviation Data Integration System, which integrates data from disparate sources into the flight data analysis process. These were incorporated into its flight operations monitoring system designed to support any aircraft and flight data recorders. The new system considers technical evolutions and needs, and each airline can perform specific treatments and build its own flight data analysis system.
page 48

Advanced Airfoils Boost Helicopter Performance

Carson Helicopters Inc. licensed the NASA-developed Langley RC4 series of airfoils to develop a replacement main rotor blade for Sikorsky S-61 helicopters. The resulting design allows the helicopters to carry heavier loads and fly faster and farther, and the main rotor blades have twice the previous service life. In aerial firefighting, the performance-boosting airfoils have helped the U.S. Department of Agriculture's Forest Service control the spread of wildfires.
page 50

Deicing System Protects General Aviation Aircraft

Kelly Aerospace Thermal Systems LLC collaborated with NASA scientists on deicing technology with assistance from the SBIR program. New and previous work combined in the development of a lightweight, easy-to-install, and reliable wing and tail deicing system called Thermawing, a DC-powered air conditioner for single-engine aircraft called Thermacool, and high-output alternators to run them both.
page 52

Public Safety

Chemical-Sensing Cables Detect Potential Threats

SBIR contracts helped Intelligent Optical Systems Inc. (IOS) develop moisture- and pH-sensitive sensors to detect corrosion or pre-corrosive conditions before significant structural damage occurs. The company subsequently worked with the U.S. Department of Defense to continue development of the sensors for detecting chemical warfare agents. IOS has also sold the technology to major automotive and aerospace companies, who are finding a variety of uses for the devices.
page 56

Infrared Imaging Sharpens View in Critical Situations

Innovative Engineering and Consulting Infrared Systems received NASA assistance to develop commercial infrared imaging systems that better differentiate the intensity of heat sources. The research resulted in the NightStalkIR and IntrudIR Alert Systems, now being used abroad to locate personnel stranded in emergency situations and protect high-value operations. The company is also applying its thermal imaging techniques to medical and pharmaceutical products.
page 58

Plants Clean Air and Water for Indoor Environments

Research begun at NASA has been continued and made publicly available by Wolverton Environmental Services Inc., including use of plants to improve indoor air quality and clean waste water. Wolverton Environmental is working with Syracuse University to tie plant-based filters into HVAC systems, has begun to assess the ability of its EcoPlanter product to remove formaldehyde from interior environments, and is in talks with designers of the new Stennis Space Center Visitor's Complex about using its designs for indoor air-quality filters.
page 60

Consumer, Home, and Recreation

Corrosive Gas Restores Artwork, Promises Myriad Applications

NASA research on corrosion and long-duration coatings led to alternate applications of atomic oxygen. Atomic oxygen was found to remove organic compounds high in carbon (such as soot) from fire-damaged artworks without altering the paint color, and has been tested on oil paintings, acrylics, watercolors, and ink. Atomic oxygen's unique characteristic of oxidizing primarily hydrogen, carbon, and hydrocarbon polymers at surface levels has also been applied to the detection of document forgeries and removal of bacterial contaminants from surgical implants.
page 64

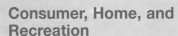

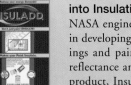

Detailed Globes Enhance Education and Recreation

Using NASA images and NOAA data, Orbis World Globes designs and produces the most visually authentic replicas of Earth ever created—EarthBalls—in many sizes that depict Earth as it is seen from space, complete with atmospheric cloud cover. Though the cloud cover has been reduced to make the landforms more visible, Orbis globes are otherwise meteorologically accurate. Applications include educational purposes from preschools to universities, games, and displays at conferences, trade shows, festivals, concerts, and parades.
page 66

Food Supplement Reduces Fat, Improves Flavor

NASA helped Diversified Services Corporation develop a nutritional fat replacement and flavor enhancement product. The now-commercialized substitute, Nutrigras, is primarily intended for use as a partial replacement for animal fat in beef patties and other high-fat meat products, and can also be used in soups, sauces, bakery items, and desserts. Nutrigras costs less than the food it replaces, and helps manufacturers reduce material costs. In precooked products, Nutrigras can increase moisture content and thereby increase product yield.
page 70

Additive Transforms Paint into Insulation

NASA engineers assisted Tech Traders Inc. in developing low-cost, highly effective coatings and paints that create useful thermal reflectance and are safe and non-toxic. The product, Insuladd, is a powder made up of inert gas-filled ceramic microspheres that is mixed into ordinary paint, allowing the paint to act like a layer of insulation. Applications include feed storage silos, poultry hatcheries, and on military vehicles and ships.
page 72

New Lubricants Protect Machines and the Environment

NASA spurred Sun Coast Chemicals to develop an effective and environmentally safe lubricant for the shuttle-bearing launcher platform. The resultant X-1R Crawler Track Lube is biodegradable and high-performance. Sensing many market opportunities, Sun Coast Chemicals introduced Train Track Lubricant, Penetrating Spray Lubricant, and Biodegradable Hydraulic Fluid; a gun lubricant/cleaner and a fishing rod and reel lubricant; and recently, the X-1R Corporation was launched to fold the high-performance, environmentally safe benefits into a line of automotive and racing products.
page 74

Environmental and Agricultural Resources

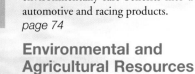

Advanced Systems Map, Monitor, and Manage Earth's Resources

A NASA Small Business Technology Transfer contract developed a hyperspectral crop-imaging project to enhance airborne hyperspectral sensing and ground-truthing means for crop inspection. SpecTIR LLC, recognized for innovative sensor design, on-demand hyperspectral data collection, and image-generating products, integrated the hyperspectral data with LIDAR systems and other commercial technologies. Areas of application include precision farming and irrigation; oil, gas, and mineral exploration; pollution and contamination monitoring; wetland and forestry characterization; water quality assessment; and submerged aquatic vegetation mapping.
page 80

Sensor Network Provides Environmental Data

The National Biocomputation Center, formed through a partnership between NASA and the Stanford University School of Medicine, focuses on telemedicine. Researchers there formed Intelesense Technologies to apply telemedicine sensors to integrated global monitoring systems to better understand links between the environment and people, monitor natural resources, predict and adapt to environmental changes, provide for sustainable development, and reduce the impacts and provide effective response to natural disasters. Current projects include tracking emerging infectious diseases such as avian influenza.
page 82

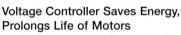

Voltage Controller Saves Energy, Prolongs Life of Motors

NASA voltage controller technology was licensed by Power Efficiency Corporation, improved, and marketed as the Performance Controller and the Power Efficiency energy-saving soft start. Soft start gradually introduces power to a motor, thus eliminating the harsh mechanical stresses of a device going from a dormant state to full activity; prevents it from running too hot; and increases the motor's lifetime. Common applications include mixers, grinders, granulators, conveyors, crushers, stamping presses, injection molders, elevators, and escalators.
page 86

Treatment Prevents Corrosion in Steel and Concrete Structures

A NASA-developed electromigration technique to prevent corrosion in rebar was combined with Surtreat Holding LLC's chemical anti-corrosive solution. NASA followed this effective match with a liquid galvanic coating for reinforced concrete, applications for which include bridge and building infrastructures, piers and docks, parking garages, cooling towers, and pipelines. Surtreat's Total Performance System, a natural compliment to the coating, provides diagnostic testing and site analysis, manufactures and prescribes site-specific solutions, controls material application, and verifies performance through follow-up testing and analysis.
page 88

Computer Technology

Optics Program Simplifies Analysis and Design

NASA engineers partnered with Midé Technology Corporation through an SBIR contract to design the Disturbance-Optics-Controls-Structures Toolbox, a software suite for performing integrated modeling for multidisciplinary analysis and design. The toolbox is being sold commercially by Nightsky Systems Inc., a spinoff company formed by Midé, to contractors developing large space-based optical systems, including Lockheed Martin Corporation, The Boeing Company, and Northrop Grumman Corporation, as well as companies providing technical audit services, like General Dynamics Corporation.
page 92

Design Application Translates 2-D Graphics to 3-D Surfaces

NASA developed a flattening process to translate surface geometry of a model to a 2-D template. Fabric Images Inc., specializing in the printing and manufacturing of fabric tension architecture for the retail, museum, and exhibit/trade show communities, utilizes software derived from NASA's to translate 2-D graphics for 3-D surfaces prior to print production. Benefits of this process include 11.5 percent time savings per project, less material wasted, and the ability to improve upon graphic techniques and offer new design services.
page 94

Hybrid Modeling Improves Health and Performance Monitoring

A NASA SBIR contract helped Scientific Monitoring Inc. create a simplified health-monitoring approach for flight vehicles and equipment. I-Trend, the resulting product, compares equipment performance to design predictions, to detect deterioration or impending failure before operation is impacted. I-Trend also characterizes health or performance alarms that result in "no fault found" false alarms. Several commercial aviation programs use I-Trend technology, and the U.S. Air Force tapped Scientific Monitoring to develop next-generation engine health-management software.
page 98

Software Sharing Enables Smarter Content Management

NASA established a technology partnership with Xerox Corporation to develop high-tech knowledge management systems and provide new tools and applications that support the Vision for Space Exploration. The first result of the partnership was the NX Knowledge Network, which combines Netmark (NASA-created practical database content management software) with complementary software from Xerox's global research centers and DocuShare. NX Knowledge Network was tested at the NASA Astrobiology Institute, and is widely used at several NASA field centers.

page 100

Engineering Software Suite Validates System Design

Five NASA SBIR contracts helped EDAptive Computing Inc. develop the EDAstar engineering software tool suite. Resulting software included Syscape, used to capture executable specifications of multi-disciplinary systems, and VectorGen, used to automatically generate tests to ensure system implementations meet specifications. Initial commercialization for EDAstar included military and defense applications, industry giants like the Lockheed Martin Corporation, Science Applications International Corporation, and Ball Aerospace and Technologies Corporation, as well as NASA field centers.

page 102

Industrial Productivity

Open-Lattice Composite Design Strengthens Structures

IsoTruss, a lightweight and efficient alternative to monocoque composite structures, developed through a series of NASA-funded projects, is garnering global attention due to its being lightweight; as much as 12 times stronger than steel; inexpensive to manufacture, transport, and install; low-maintenance; and fully recyclable. Expected applications include utility poles and meteorological towers, concrete structures, sign supports, prostheses, irrigation equipment, and sporting goods.

page 106

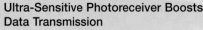

Ultra-Sensitive Photoreceiver Boosts Data Transmission

Epitaxial Technologies LLC was awarded a NASA SBIR contract to address advanced sensor needs. The resulting photoreceiver is capable of single proton sensitivity and is smaller, lighter, and requires less power than its predecessor. The receiver operates in several wavelength ranges; will allow data rate transmissions in the terabit range; and will enhance Earth-based missions for remote sensing of crops and other natural resources. Widespread military and civilian applications are anticipated.

page 108

Micro Machining Enhances Precision Fabrication

Two NASA SBIR contracts helped Creare Inc. develop an ammonia evaporator for thermal management systems. Founded to commercialize this work, Mikros Technologies Inc. developed advanced micro-electrical discharge machining to produce tiny holes in the evaporator. This technique was used to fabricate micro-nozzle array systems for industrial ink jet printing systems. Mikros, now the world leader in fabricating stainless steel micro-nozzles, was awarded two NASA SBIR contracts to advance micro-fabrication and high-performance thermal management technologies.

page 110

Portable Hyperspectral Imaging Broadens Sensing Horizons

Photon Industries Inc., a spinoff of NASA and the Institute for Technology Development dedicated to developing hyperspectral imaging technologies, was purchased by Lextel Intelligence Systems LLC. Lextel added new features and expanded the applicability of the hyperspectral imaging systems, making advances in size, usability, and cost. The company now offers a suite of turnkey hyperspectral imaging systems, based on the original NASA groundwork, used worldwide for a wide variety of applications including medical, military, forensics, and food safety.

page 112

Hypersonic Composites Resist Extreme Heat and Stress

Research contracts with NASA helped Materials and Electrochemical Research Corporation (MER) develop technologies for hypersonic flights, including a coating that passed simulated Mach 10 conditions and carbon-carbon (C-C) composite components. The C-C composites are very lightweight and exceptionally strong, even at very high temperatures. MER formed Frontier Materials Corporation to introduce these materials to the commercial markets. The composites have been used in industrial heating applications, in the automotive and aerospace industries, in glass manufacturing, and on semiconductors.
page 114

Computational Modeling Develops Ultra-Hard Steel

QuesTek Innovations LLC developed a carburized, martensitic gear steel with an ultra-hard case using proprietary computational design methodology. NASA researchers conducted spur gear fatigue testing for the company with a spiral bevel or face gear test rig, which revealed that QuesTek's gear steel outperforms the current state-of-the-art alloys used for aviation gears in contact fatigue by almost 300 percent. Uses for this new class of steel are limitless in areas needing exceptional strength for high-throughput applications.
page 116

Thin, Light, Flexible Heaters Save Time and Energy

EGC Enterprises Inc. used NASA's Icing Research Tunnel to develop thermoelectric thin-film heater technology to address in-flight icing on aircraft wings. Working with NASA researchers and the original equipment manufacturers of aircraft parts, the company developed the Q•Foil Rapid Response Thin-Film Heater. The product meets all criteria for in-flight use and promises a broad range of applications, including cooking griddles, small cabinet heaters, and several laboratory uses.
page 117

Novel Nanotube Manufacturing Streamlines Production

The NASA Innovative Partnerships Program promoted a NASA-developed process for creating nanotubes, and Idaho Space Materials Inc. (ISM) applied for a nonexclusive license for the single-walled carbon nanotube manufacturing technology. ISM commercialized its products, and the inexpensive, robust nanotubes are now being used to create the next generation of composite polymers, metals, and ceramics. Researchers are also examining ways to use these materials in myriad technologies, including transistors, fuel cells, televisions, supercapacitors, catalysts, and advanced composite materials.
page 118

'NASA Invention of the Year' Controls Noise and Vibration

NASA's Macro-Fiber Composite (MFC) is designed to control vibration, noise, and deflections in composite structural beams and panels. Smart Material Corporation licensed the MFC technology to add it to their line of commercially produced actuators, and to date has sold MFCs to over 120 customers, including Volkswagen, Toyota, Honda, BMW, General Electric, and the tennis company, HEAD. Consumer applications already on the market include audio speakers, phonograph cartridges, microphones, and products requiring vibration control such as sports equipment.
page 120

Thermoelectric Devices Advance Thermal Management

United States Thermoelectric Consortium Inc. (USTC), in working to integrate the benefits of thermoelectric devices in its line of thermal management solutions, has found NASA technical research to be a valuable resource. In cooperation with NASA, USTC built a gas emissions analyzer (GEA) for combustion research which precipitated hydrocarbon particles, preventing contamination that would hinder precise rocket fuel analysis. USTC work has since provided thermal solutions for computer, radar, laser, microwave, and other systems.
page 122

NASA Technologies Enhance Our Lives

International Space Station

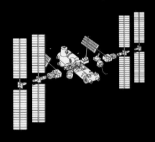

Space Telescopes and Deep Space Exploration

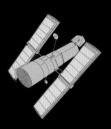

Satellites and Imaging Technology

Innovative technologies from NASA's space and aeronautics missions (above) transfer as benefits to many sectors of society (below).

Each benefit featured in *Spinoff* 2007 is listed with an icon that corresponds to the mission from which the technology originated.

Health and Medicine

Improves CPR
page 22

Detects cardiovascular disease
page 26

Assists patients with cognitive disorders
page 30

Evaluates nerve function
page 32

Fights acne
page 34

Broadens cellular analysis
page 36

Enhances diagnostic Imaging
page 40

Transportation

Eases air traffic management
page 44

Advances rotorcraft design
page 46

Improves flight safety
page 48

Boosts helicopter performance
page 50

Protects general aviation aircraft
page 52

Public Safety

Detects potential threats
page 56

Sharpens views in critical situations
page 58

Cleans air and water for indoor environments
page 60

Consumer, Home, and Recreation

Restores artwork
page 64

Enhances education and recreation
page 66

Reduces fat while improving flavor
page 70

Transforms paint into insulation
page 72

Protects machines and the environment
page 74

Environmental and Agricultural Resources

Maps, monitors, and manages Earth's resources
page 80

Provides environmental data
page 82

Saves energy and prolongs motor life
page 86

Prevents corrosion in steel and concrete structures
page 88

Computer Technology

Simplifies analysis and design
page 92

Translates 2-D graphics to 3-D surfaces
page 94

Improves health and performance monitoring
page 98

Enables smarter content management
page 100

Validates system design
page 102

Industrial Productivity

Strengthens structures
page 106

Boosts data transmission
page 108

Enhances precision fabrication
page 110

Broadens sensing horizons
page 112

Resists extreme heat and stress
page 114

Develops ultra-hard steel
page 116

Saves time and energy
page 117

Streamlines production
page 118

Controls noise and vibration
page 120

Advances thermal management
page 122

NASA Technologies Benefiting Society

The National Aeronautics and Space Act of 1958 required that NASA disseminate its information to the public, and the Technology Utilization Act of 1962 formalized the process through which the Agency was to accomplish this task. Today, NASA continues to seek industry partnerships to develop technologies that apply to NASA mission needs, provide direct societal benefits, and contribute to competitiveness in global markets. As part of NASA's mission, the Agency facilitates the transfer and commercialization of NASA-sponsored research and technology. These efforts not only support NASA, they enhance the quality of life in our hospitals, homes, and communities.

Health and Medicine

NASA research drives innovation that improves health care. The technologies featured in this section:

- Improve CPR
- Detect cardiovascular disease
- Assist patients with cognitive disorders
- Evaluate nerve function
- Fight acne
- Broaden cellular analysis
- Enhance diagnostic imaging

Circulation-Enhancing Device Improves CPR

Originating Technology/NASA Contribution

Ever stand up too quickly from a sitting or lying position and feel dizzy or disoriented for a brief moment? The downward push of Earth's gravity naturally causes blood to settle in the lower areas of the human body, and occasionally, with a quick movement—such as rising swiftly from a chair—the body is not able to adjust fast enough to deliver an adequate supply of blood to the upper parts of the body and the brain. This sudden, temporary drop in blood pressure is what causes brief feelings of lightheadedness upon standing. In essence, when the heart pumps blood to different parts of the body, it is working against the physical phenomenon of gravity in its efforts to send blood up to the brain.

In more cases than not, the body is able to make the necessary adjustments to ensure proper blood flow and pressure to the brain; but when the disorientation lasts a long time and/or become chronic, individuals may have a condition called orthostatic intolerance. According to the American Journal of Physiology–Heart and Circulatory Physiology, an estimated 500,000 Americans are affected by orthostatic intolerance. Symptoms range from occasional fainting, blurry vision, and pain or discomfort in the head and the neck, to tiredness, weakness, and a lack of concentration. Though research indicates that the condition is not life-threatening, it could impact the quality of life and contribute to falls that result in serious injuries.

The condition is a prominent concern for NASA, since astronauts have to readjust to the gravitational environment of Earth after spending days in the weightlessness of space. NASA's Exploration Systems Mission Directorate has found that roughly 20 percent of astronauts coming off of short-duration space flights experience difficulty maintaining proper blood pressure when moving from lying down to either sitting or standing during the first few days back on Earth. The difficulties are even more severe for astronauts coming off of long-duration missions, according to the mission directorate, as 83 percent of these crewmembers experience some degree of the condition.

Cardiovascular experts at NASA have found that the blood that normally settles in the lower regions of the body is instead pulled to the upper body in the microgravity environment of space. Blood volume is subsequently reduced as some cardiovascular reflexes are no longer being used, and less blood flows to the legs. Additionally, the muscles weaken, especially in the lower portion of the body, because they are not working (contracting) as hard as they usually do. This is not so much a concern for the astronauts while they are in space, since the action of floating around takes the place of putting center-of-gravity pressure on their legs. (They do exercise strenuously while in microgravity, though, to keep their muscles and circulatory systems conditioned, thus preparing their bodies for the return to gravity as best they can.) When they return to Earth's gravity, however, more blood returns to the legs. Since there is a lower volume of blood, the flow that is supposed to be traveling to the brain can be insufficient. That is when orthostatic intolerance can set in.

NASA has conducted and sponsored a wealth of studies to counter the effects of orthostatic intolerance, especially since the condition could prevent an astronaut from exiting a landed spacecraft in the event of an emergency. In one study conducted by Johnson Space Center's Cardiovascular Laboratory, astronauts in orbit tested the efficacy of a drug called midodrine that has successfully reduced orthostatic intolerance in patients on Earth. The early results were promising, but further testing will be conducted by the laboratory before more conclusive results can be determined. In another study, the laboratory is using a controlled tilt test on Earth to replicate the body's responses to a shift from reclining to sitting or standing.

At Ames Research Center, researchers are utilizing NASA's 20-G artificial gravity centrifuge machine in a pilot study on cardiovascular responses and fluid shifts in the body. A separate Ames study is evaluating the possibility of expanding astronauts' plasma volumes (the fluid part of the blood, minus the blood cells), as a preventative measure.

The ResQPOD is an impedance threshold device used to enhance circulation during CPR. It could be used to increase circulation for astronauts as their bodies initially adjust to a return to gravity from the weightlessness of space.

In NASA-sponsored research at Vanderbilt University, researchers have successfully identified a genetic cause for orthostatic intolerance. The findings marked the first time a genetic defect had been linked to a disorder of the autonomic immune system, according to the discoverers, and could eventually lead to new drugs and treatments for the condition.

At Kennedy Space Center, a collaborative research effort with the U.S. Army and private industry has yielded an important application for a new, non-invasive medical device called ResQPOD that is now available for astronauts returning from space. In helping to reacquaint the astronauts with the feeling of gravity, ResQPOD quickly and effectively increases the circulation of blood flow to the brain. This device is also available to the public as a means to enhance circulation for breathing patients suffering from orthostatic intolerance and for non-breathing patients suffering cardiac arrest or other high-risk clinical conditions attributed to low blood pressure.

Partnership

Advanced Circulatory Systems Inc., of Minneapolis, collaborated with Kennedy and the U.S. Army Institute of Surgical Research for more than 5 years to develop ResQPOD. Don Doerr, an engineer at Kennedy, led the testing and development effort; Dr. Victor Convertino of the Institute of Surgical Research (and a former NASA scientist at Kennedy) also played an instrumental role in developing the technology.

Multiple clinical studies were conducted during the research effort, including six published studies. The published works demonstrate that ResQPOD offers a significant improvement in cardiac output and blood flow to the brain and in preventing shock in the event of considerable blood loss, when compared to conventional resuscitation. According to Advanced Circulatory Systems, data from the NASA studies played a major role in the company obtaining U.S. Food and Drug Administration 501K clearance for the device.

The ResQPOD increases circulation in states of low blood pressure. When used on patients in cardiac arrest, the ResQPOD harnesses the chest wall recoil after each compression to generate a small but critical vacuum within the chest. This vacuum enhances blood flow back to the heart and results in a marked increase in blood flow out of the heart with each subsequent chest compression.

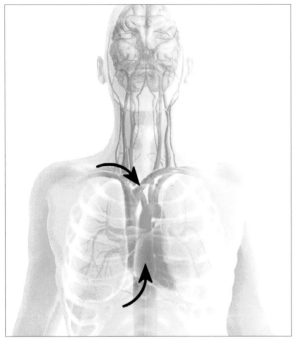

CPR delivers approximately 15 percent of normal blood flow to the heart.

The ResQPOD doubles blood flow back to the heart.

During the decompression (release) phase of CPR, an increase in negative pressure in the thoracic cavity results in drawing more blood back into the chest, providing greater venous return to the heart.

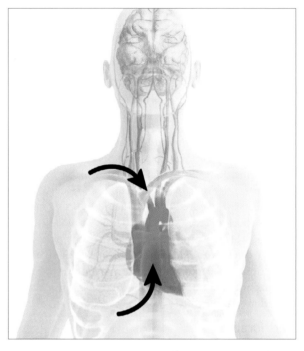

Dr. Keith Lurie, chief medical officer at Advanced Circulatory Systems and a primary member of the collaborative research effort, said, "The three-way partnership between NASA, private industry, and the U.S. Army Institute of Surgical Research is really a model for how organizations can work together to benefit both government programs and civilians."

In 2006, Dr. Smith Johnston, the lead flight surgeon for NASA's space shuttle missions, added ResQPOD to the list of medical equipment that is available for returning astronaut crews. The device was on hand for the landing of Space Shuttle Atlantis (STS-115) on September 21, 2006.

"We're excited that our devices were available to the medical team [for the STS-115 mission] and look forward to continued collaboration with NASA to assist

its efforts to safeguard the health of the astronauts," added Lurie.

Product Outcome

Manufactured commercially by Advanced Circulatory Systems and distributed by Sylmar, California-based Tri-anim Health Services Inc., the ResQPOD circulatory enhancer improves upon the standard of care for patients with a variety of clinical conditions associated with low blood flow. Advanced Circulatory Systems' primary commercial focus, though, is on non-breathing patients who can benefit from enhanced circulation, such as those experiencing cardiac arrest.

According to the American Heart Association, about 900 Americans fall victim to sudden cardiac arrest every day, with approximately 95 percent dying before they reach the hospital. This is why cardiopulmonary resus-

citation (CPR) can mean the difference between life and death, as increasing blood flow to the heart and brain until the heart can be restarted is critical to improving survival rates with normal neurological functioning.

ResQPOD is an American Heart Association-rated Class IIa impedance threshold device, meaning that it is the highest recommended "adjunct" in the association's latest guidelines for CPR. As a Class IIa impedance threshold device, it also carries a higher recommendation than any medication used to boost circulation in adults suffering cardiac arrest, according to these guidelines.

The device is about the size of a fist and can be affixed to either a facemask or an endotracheal breathing tube during CPR. It enhances the intrathoracic vacuum that forms in the chest during the chest recoil phase of CPR by temporarily sealing off the airway between breaths and preventing unnecessary air from entering the chest

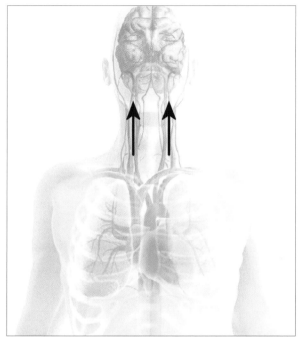

CPR delivers approximately 25 percent of normal blood flow to the brain.

The ResQPOD delivers more than 70 percent of normal blood flow to the brain.

Improved venous return results in increased cardiac output during the subsequent compression phase of CPR, providing greater blood flow to the brain.

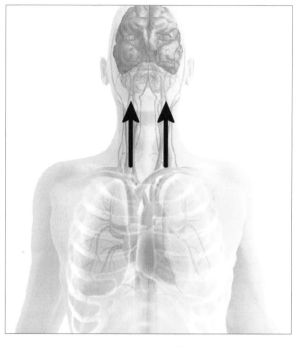

(timing-assist lights on the device will aid the rescuer in ventilating the patient at a proper rate). The vacuum that is created pulls blood back to the heart, doubling the amount of blood that is pulled back by conventional mouth-to-mouth/chest compression CPR, according to clinical studies, which also show that blood flow to the brain is increased by 50 percent. In sustaining proper blood flow to the heart and to the brain, ResQPOD increases the likelihood of survival and decreases the likelihood of neurological disorders.

ResQPOD is being used by emergency medical technicians in cities all around the country, including Boston, Houston, Indianapolis, Miami, and Oklahoma City, as well as Hartford, Connecticut; Kansas City, Missouri; Raleigh, North Carolina; and Toledo, Ohio. In some cities, it has reportedly increased the number of cardiac arrest patients delivered alive to the hospital by as much

as 50 percent. At Cypress Creek Emergency Medical Services (EMS), a large medical care organization serving more than 400,000 residents in the greater Houston area, ResQPOD has become a standard of care. Overall resuscitation rates climbed to nearly 50 percent since the organization began deploying the device in 2005, boosting hospital admission rates from 26 percent to an astounding 38 percent.

"These results are gratifying, and we applaud the entire Cypress Creek EMS organization for their advanced emergency medical service care and their ability to turn around the dismal statistics that surround cardiac arrest," noted Advanced Circulatory Systems' Lurie.

In its secondary commercial applications, Advanced Circulatory Systems is offering ResQPOD to improve circulation in patients suffering from orthostatic intolerance and general low blood pressure. These secondary uses also

apply to individuals who undergo dialysis treatments and may experience a drop in blood pressure, as well as those who go into shock after severe blood loss.

Outside of the traditional hospital setting, the company is investigating the beneficial impact ResQPOD could have on wounded soldiers in the battlefield who may have lost a great deal of blood and are in danger of going into shock.

Advanced Circulatory Systems is also harnessing the physiological principles discovered during its research collaboration with NASA to develop another promising technology: an intrathoracic pressure regulator for patients requiring ventilation assistance because they are too sick to breathe on their own. ❖

ResQPOD® is a registered trademark of Advanced Circulatory Systems Inc.

Noninvasive Test Detects Cardiovascular Disease

Originating Technology/NASA Contribution

For decades now, NASA has been sending spacecraft throughout the galaxy. Once in the cosmos, these crafts use advanced cameras to create images of corners and crevices of our universe never before seen and then transmit these pictures back to laboratories on Earth, where government scientists then ask themselves: *What exactly are we looking at?*

That question is answered at NASA's Jet Propulsion Laboratory (JPL) in the Image Processing Laboratory, founded in 1966 to receive and make sense of spacecraft imagery. There, NASA-invented VICAR (Video Image Communication and Retrieval) software has, through the years, laid the groundwork for understanding images of all kinds. The original software, created by a JPL team of three, Robert Nathan, Fred Billingsley, and Robert Selzer, is in use even to this day, although with greater accuracy and effectiveness due to decades of advancements.

The imaging division at JPL has grown increasingly sophisticated over the years, developing new processes and technologies to handle increasingly complex acquisitions from each NASA space mission, from the Voyager images of Saturn and Jupiter taken in the 1970s, to the new imagery captured by the Mars Reconnaissance Orbiter in late 2006 that suggests water still flows on Mars, opening the possibility that the Red Planet could perhaps support some forms of life.

Partnership

Selzer, from the original VICAR team, has made the NASA imaging technology his life's work, spending 46 years as a NASA employee and continuing to work on its advancement even after his retirement from JPL. Selzer received many NASA awards for technical achievement, including the prestigious Technology and Application Program Exceptional Achievement Medal.

During the last 15 years of his career as a government scientist, he, as head of the JPL Biomedical Image Processing Laboratory, was working on using the imaging technology for health care diagnosis.

The project began when the imaging team developed the idea of using the VICAR software to analyze X-ray images of soft tissue. Typically, the X-ray is ineffective when used to analyze soft tissues, though the researchers were curious to see if the imaging software could broaden the application of this readily available diagnostic procedure. The results were interesting, but too much quality was lost in transferring the pictures into a digital format for analysis. Still, the idea seemed feasible, so, with several grants from NASA, the testing continued.

Selzer's JPL team, partnering with scientists from the University of Southern California under the direction of the late Dr. David Blankenhorn and Dr. Howard Hodis, director of the Atherosclerosis Research Unit at the school's Keck School of Medicine, began to image X-rays of arteries. With marginal success using X-rays, they came upon the idea of using the same methodology but applying it to ultrasound imagery, which was already digitally formatted. The new approach proved successful for assessing amounts of plaque buildup and arterial wall thickness, direct predictors of heart disease.

Testing continued, and the team, buoyed by its successes, began looking for outside funding and methods

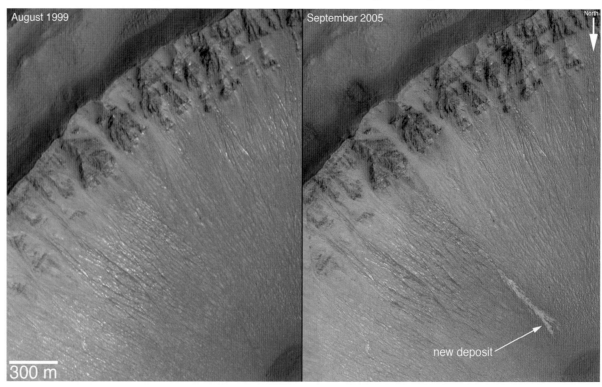

Images from the Mars Reconnaissance Orbiter suggest that liquid water has flowed on Mars within the past 6 years. These pictures were analyzed and clarified with imaging software technologies from NASA's Jet Propulsion Laboratory.

of distribution. At this point, Gary F. Thompson entered the picture.

Thompson has a history of heart disease in his family. The first male in many generations to live past age 50, and the last living male in his family line, Thompson comes from a long line of active, athletic men who, with no prior symptoms, suffered fatal heart-related events. A lifelong athlete who had boxed in the New York City Golden Gloves tournament in his youth and ran his first marathon in 1975, Thompson was understandably concerned, but feeling confident, when he approached his 50th birthday. He had the family history working against him, but he was also in prime shape. To celebrate his half-century mark, he planned to run three marathons: Los Angeles, New York, and Boston, and he underwent a battery of medical tests, all of which confirmed that he was in perfect health, without any signs of cardiovascular disease.

Seven days after his birthday, Thompson ran the Los Angeles marathon. At the 15th mile, he started experiencing back pain. By the 20th mile, it became so excruciating, that he stopped running and sought the help of a police officer who was monitoring the race. Later, at the hospital, doctors confirmed that he had suffered a moderate heart attack and lost 48 percent of his heart muscle. The modern medical testing, he realized, had failed him. Luckily for Thompson, compared to most men his age, for him to have 52 percent of his heart working was the equivalent of 127 percent, because he was so athletic.

Months later, at dinner with David Baltimore, then president of the California Institute of Technology (Caltech), Thompson asked whether there were any new heart-related breakthroughs from the esteemed university and was surprised to hear that there actually was, indeed, a new technology, but that it had been developed at JPL. It was a noninvasive diagnostic system with the ability to accurately predict heart health. Baltimore offered to set up an appointment for Thompson at the University of Southern California hospital where this new

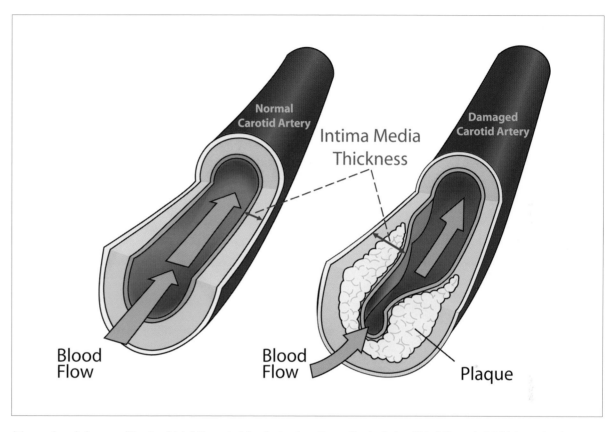

Atherosclerosis is a condition in which fatty material collects along the walls of arteries. This fatty material thickens, hardens, and may eventually block the arteries.

method was being tested, but Thompson declined the privileged treatment.

Instead, he went in on his own, unannounced, and without revealing his family history or recent heart attack. Thompson, who, despite the recent event, gave every impression of great health, met with a technician, who asked him to loosen his collar and then performed an ultrasound scan on either side of his neck, the location of the carotid arteries. When the results were in, the technician told Thompson that he needed to meet with the doctor immediately. The test showed that Thompson had

the thickest artery walls of the over 3,000 other people tested, a direct indicator that he was in danger of a heart attack or stroke.

Thompson was impressed. This new device had managed to do what all of the other tests had failed to do: give him an accurate reading of his heart health. Thompson, a hard-charging entrepreneur, met with the researchers Selzer and Hodis and told them that they needed to get this technology into the hands of physicians. They agreed. Thompson developed a business plan, secured an exclusive license for the JPL-developed technology from

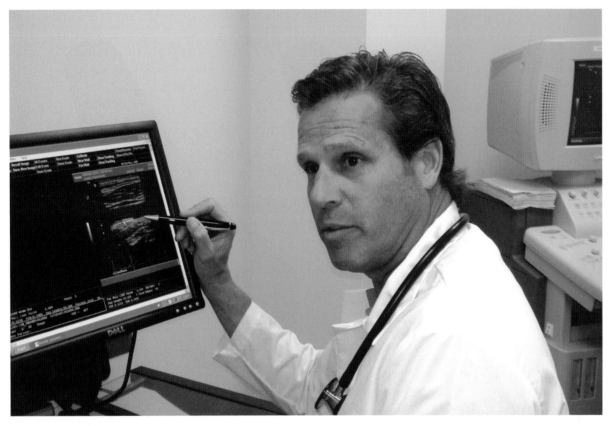

and fatty substances in the arteries, combined with arterial hardening. The result is that blood flow through the heart is restricted, hampering oxygen supply. Heart attacks occur when the heart does not get the necessary oxygen, and strokes are the result of oxygen not reaching the brain. Atherosclerosis, referred to by the American Heart Association (AHA) as the "silent killer," initially has no discernable symptoms until one or more of the arteries becomes so congested that these major, sometimes fatal, problems occur. The AHA estimates that two out of three unexpected cardiac deaths occur without prior symptoms.

In fact, astronaut Edward White, the first astronaut to ever perform a space walk, and one of the three space pioneers to die during the Apollo launch pad tragedy in 1967, was thought by most to be in perfect health, having successfully passed all of the rigorous astronaut testing. An autopsy directly after the accident, though, revealed that he had extreme thickening of the arteries and showed most signs of heart disease. With this particular disease, though, as has been mentioned previously, often the only symptoms are either heart attack or stroke.

ArterioVision provides a direct measurement of atherosclerosis before it causes these "symptoms" by safely and painlessly measuring the thickness of the first two layers of the carotid artery wall using an ultrasound procedure and advanced image-analysis software. The carotid artery, located on both sides of the neck, is the largest artery that is near the surface of the skin. It supplies blood to the brain, and it can therefore be examined noninvasively. ArterioVision essentially performs a skin surface imaging "biopsy" to examine the arterial wall.

Caltech, and invested his own money to start Medical Technologies International Inc. (MTI), located in Palm Desert, California. Selzer, after retiring from JPL, joined in and now serves as the company's chief engineer.

Product Outcome

MTI licensed 14 research institutions around the world for pre-U.S. Food and Drug Administration (FDA) clearance, research-only use of the analysis software and then incorporated feedback from these groups into the new clinical product it was developing. It patented the new developments and then submitted the technology to a rigorous review process at the FDA, which cleared the device for public use. MTI also filed with the American Medical Association to have the device given a dedicated Current Procedural Technology (CPT) code for insurance purposes, thus encouraging more doctors to offer this test to patients.

The patented software is being used in MTI's ArterioVision, a carotid intima-media thickness (CIMT) test that uses ultrasound image-capturing and analysis software to noninvasively identify the risk for the major cause of heart attack and strokes: atherosclerosis.

The term atherosclerosis comes from Greek, with athero meaning a gruel or paste (plaque) and sclerosis meaning hardness. It is just that, a buildup of cholesterol

As evidenced by the battery of tests Thompson underwent before his heart attack, diagnostic tools for atherosclerosis screening are far from advanced. Atherosclerosis begins in the abdomen and ascends to the heart and the carotid arteries. A diagnostic tool for examining it in the abdomen is the computerized tomography (CT) scan, but CT comes with a certain level of risk and at a great financial cost. Unlike ultrasound, which is safe enough that it is used on unborn babies, CT scans rely on radiation to produce results. While this procedure is still new enough that the risks are relatively unknown, the National Toxicology Program, a division of the U.S. Department of Health and Human Services, has recently declared X-rays carcinogenic. Similarly, researchers at Columbia University have estimated that radiation for a full-body scan is roughly equivalent to the same amount of radiation exposure experienced by people within 1.5 miles of the atomic bombs dropped on Nagasaki and Hiroshima. In other terms, it is roughly equivalent to the radiation from 200 chest X-rays. As an added disadvantage, X-ray machines and the production of radiation cost a great deal of money.

The NASA-based technology has none of these problems. It poses no risks, and it is relatively inexpensive. The imaging technology can distinguish between 256 different shades of gray and differentiate nuances at a sub-pixel level of interpolation, making it the most accurate in this field, and it is compatible with all existing ultrasound machines, making it more readily accessible to physicians.

While ArterioVision is not the only FDA-cleared CIMT tool on the market, it is the only one that offers a predictive report for the physician and patient. It explains the significance of test results using a proprietary database and JPL-developed algorithms, and can extrapolate to show percentile of risk.

One particular feature of the report is the revelation of arterial age. It can show the patient that while he may be 50 years old, his arteries may be the equivalent of patients 75 years old. This real-world number is something

patients can identify with and helps promote compliance with drug therapies and other forms of treatment—one of the most difficult aspects of preventing and treating heart disease. Physicians often lament that they stress the importance of lifestyle changes to their patients, but since heart disease does not initially come with symptoms, but instead with potentially fatal events, it is often difficult to impress upon the patients the urgency of taking care of their hearts. The ArterioVision patient report provides a significant warning sign and gives concrete examples.

The report then becomes part of a serial examination. Atherosclerosis can be reversed with strategies like exercise, diet change, weight loss, smoking cessation, and cholesterol-lowering medication; and the patient can see, after implementing a new strategy, if that one is working.

Currently, the technology is in all 50 states, and in many countries throughout the world. MTI is continuing to push this life-saving technology and is rapidly expanding its sales force in an effort to live up to its company credo, "Making a Positive Difference Every Day." ❖

ArterioVision™ is a trademark of Medical Technologies International Inc.

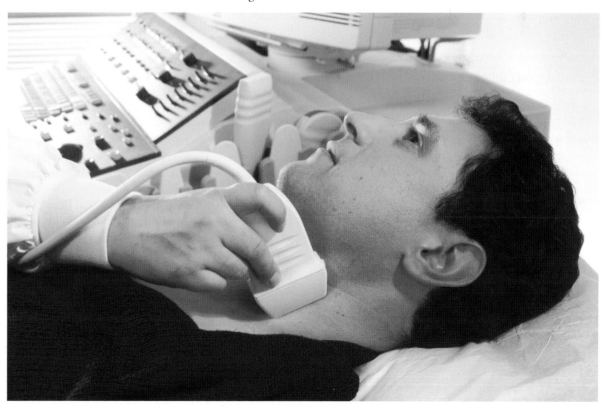

ArterioVision is a CIMT test that uses ultrasound image-capturing and analysis software to noninvasively identify the risk for the major cause of heart attack and strokes: atherosclerosis.

Scheduling Accessory Assists Patients with Cognitive Disorders

Originating Technology/NASA Contribution

Neuropsychology is the study of how the brain relates to behavior, emotion, and cognition. Clinical neuropsychologists evaluate the behavioral effects of neurological and developmental disorders stemming from brain injury, strokes, multiple sclerosis, Alzheimer's disease, and Parkinson's disease. Millions of Americans are currently living with these cognitive disorders, including a growing number of veterans returning from Iraq with brain injuries. The disorders often result in cognitive impairments which make it difficult to plan daily activities and stay on task, affecting independence, quality of life, and employment.

In the early 1990s, Richard Levinson, a NASA contractor and senior researcher in Ames Research Center's Artificial Intelligence Research Branch and Autonomous Systems Group, took the science in an entirely different direction when he folded it into his NASA work.

Levinson, who had previously received a Space Act Award for contributing to the development of a prototype autonomous thermal control system for the International Space Station, initially learned about neuropsychology accidentally. Moving into a new apartment in 1986, he received a neuropsychology course program in the mail that was actually intended for the previous occupant. The topic intrigued Levinson, and the pursuit to learn more was on.

As Levinson learned about emerging neuropsychological models of human planning, he continued researching computer models of automatic planning. A central concern for both fields is that plans often change when surprises occur. Neuropyschologists study how this integrated planning and execution breaks down as the result of cognitive impairment, but they do not know exactly how planning occurs in the brain. On the other side, computer scientists can build a planning system, but have a limited understanding of how to integrate planning with execution monitoring and error recovery.

Levinson studied the neuropsychology of human planning and applied this knowledge to his NASA research in order to increase autonomy for spacecraft and robots. Since spacecraft and robots operate in uncertain conditions, they cannot be preprogrammed for every activity, so there are times that they must be responsible for their own "health" and safety. Further, as NASA's missions grow more complex, so does the Agency's need for machines that can exhibit a higher degree of independence and execute improvised actions in novel situations where preprogrammed commands will not work.

In 1995, Levinson published peer-reviewed research papers in computer science and neuropsychology journals, describing an artificially intelligent planning and reaction model founded on neuropsychological theories of human behavior. This planning and reaction model was based on the functioning of the human brain's frontal lobes, which play a part in memory, motor skills, planning, decision making, and socialization, among other functions.

Levinson has since received three patents for the technology, pertaining to activity planning and cueing methods with execution monitoring and error correction.

While Levinson and NASA continue to investigate this advanced computer model for future missions, the technology has already made its terrestrial debut in the form of a powerful cueing and scheduling aid to help people with a wide range of cognitive, attention, and developmental disorders.

Partnership

Levinson received initial funding from NASA and his contracting company, Recom Technologies Inc., of Roseville, California, to research the commercial potential of his artificially intelligent planning reaction model to serve as a tool for helping individuals suffering from various forms and levels of brain impairment. In 1993, the chief of Ames' Artificial Intelligence Research Branch suggested that Levinson contact Santa Clara Valley Medical Center, which hosts a nationally acclaimed rehabilitation and research center that specializes in brain injuries, to see if the hospital was interested in a research collaboration. Levinson heeded the advice and found a valuable partner in the medical center. This partnership led to further development of Levinson's technology and funding to support clinical research from the U.S. Department of Education's National Institute on Disability and Rehabilitation Research.

In 1996, Levinson founded Attention Control Systems Inc., in Mountain View, California, to produce and market this NASA spinoff creation.

Product Outcome

Attention Control Systems now offers people with memory, attention, and cognitive disorders a computerized, personal planning device to help them stay on task by overcoming limitations in planning and fulfilling their daily schedules. The device, called the Planning and Execution Assistant and Trainer, or PEAT, is a pocket-sized PDA, complete with a graphical display, touchscreen controls, an electronic calendar, an address book, and a built-in phone. The functionality of PEAT, however, transcends that of a regular PDA scheduling device. PEAT cues users to start or stop scheduled activities, monitors their progress, and adjusts schedules as necessary in response to delays or calendar changes. It uses the automatic planning model developed for NASA to make automatic adjustments to daily plans when a situation changes. Most PDA systems lack this flexibility, requiring their users to manually re-plan and update schedule data when changes occur.

While daily routine activities come naturally to most, individuals with memory, attention, and cognitive impairment may struggle to remember that they have to perform certain tasks. Those with severe impairment to the point where independent living is a challenge are affected most, as they may not only forget to perform tasks, but forget how to perform them.

Whether individuals are mildly or severely impaired, PEAT makes it easier for them to get through their planned schedules by providing cues for task completion and adjusting for unplanned schedule conflicts. PEAT can automatically shift flexible tasks that do not require an exact start time in order to keep the prioritized, scheduled events on track. For example, an individual using PEAT wakes up to a preplanned day that consists of having breakfast with a family member from 9:30 to 10:30, followed by stopping at the bank, and then seeing a 12:00 matinee show with a friend (the individual receives cues from PEAT to inform him/her of all of these scheduled tasks). This agenda was preprogrammed in the user's device (either programmed by the user or by a caregiver, depending on the degree of impairment), with breakfast and the movie being the top-priority scheduled tasks, and the bank trip being a secondary, unscheduled routine task.

Not everything goes as planned, however. It turns out that breakfast takes longer than the scheduled hour, so the user does not have time to stop at the bank before the movie. Since the bank trip did not require an exact start time, it is a task that PEAT can automatically shift to another available time. This way, the task, though delayed, is not ignored and will not be forgotten, and the individual's priority tasks—breakfast and the movie—are not interrupted.

The automatic cues that PEAT delivers to its users to start and stop activities can be in the form of customized voice recordings, sounds, and pictures; extra large text and pictures help users with visual and motor problems. Cueing continues until the user responds. Additionally, users can program customized scripts (activity sequences) for breaking large tasks into multiple, small tasks. This feature is especially helpful for highly impaired users who may find difficulty completing tasks such as getting dressed in the morning or fixing themselves a meal.

PEAT's Cue Card display provides a countdown timer until the next scheduled event and cues the user to start

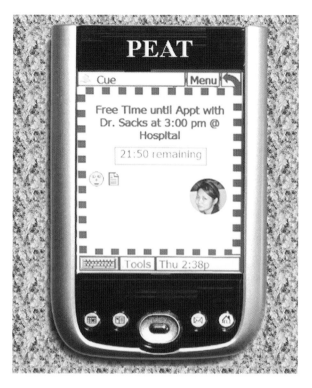

PEAT is a powerful cueing and scheduling aid to help people with a wide range of cognitive, attention, and developmental disorders.

or stop at the scheduled time. Highly impaired users may be locked into this Cue Card section, which means they will only see reminders for one event at a time. This mode keeps the reminders simple and does not create unwarranted confusion for these users. Other users with less impairment may have the option to override cues by starting and stopping them early, and can delay, skip, reschedule, or altogether cancel the cues. The device keeps a log of all of these actions, so that the caregivers and family members can evaluate how well a user is adapting to the technology and accomplishing real-world tasks. For the user, the accomplishments boost independence and confidence, and diminishes cost of care.

"PEAT can be simplified by hiding features so that highly impaired users will use a system with far less features than higher-functioning users," stated Levinson. "We start off simple with each user and add options over time. In some cases, a caregiver or therapist sets up the schedule and the user must only respond to cues, while PEAT monitors their progress and automatically adjusts the schedule as necessary," he added.

PEAT is sold as a complete system that includes software, hardware, documentation, and technical support. In addition to the flagship Pocket PEAT device, there is PEAT Phone: software that runs on cellular phones; PC PEAT: software that runs on desktop and laptop PCs, where the larger screen and keyboard may be used to configure the system, enter data, train users, and back up data; and PEAT Link: software that links the Pocket PEAT device to PC PEAT for software copying and data transfer.

PEAT is currently providing planning and execution assistance to patients at Department of Rehabilitation facilities in 25 states, Santa Clara Valley Medical Center, the U.S. Department of Veterans Affairs' Palo Alto Health Care System, and to school districts and assistive technology centers.

At the Palo Alto hospital's Polytrauma Rehabilitation Center, Dr. Harriet Zeiner, lead clinical neuropsychologist, has developed treatment protocols for troops returning from overseas with mild traumatic brain injuries from improvised explosive devices, as well as for soldiers with post-traumatic stress disorder. Zeiner's treatments include using the PEAT device as a memory prosthesis.

Meanwhile, clinical studies of PEAT continue at Santa Clara Valley Medical Center. Levinson also foresees the technology he first developed for autonomous robotic planning to have "spin-in" application for NASA's astronauts. ❖

PEAT™ is a trademark of Attention Control Systems Inc.

Neurospinal Screening Evaluates Nerve Function

Originating Technology/NASA Contribution

When in the zero-gravity environment of space, an astronaut realizes quickly that most motions require significantly less effort, and the body adjusts itself to the new environment so that a simple act like putting in a contact lens does not result in a sharp poke in the eye or clapping of hands does not shatter fingers. This adaptability is useful and necessary while in orbit, and the body quickly becomes accustomed to the zero-gravity conditions of space flight, but without the everyday weight of gravity that we often take for granted providing resistance, muscle tissue tends to atrophy. In fact, a space traveler often experiences a feeling of heaviness, of an additional weight on the body, upon returning from space. The condition is similar to the degeneration of muscle seen in bedridden patients and the elderly.

Naturally, NASA is invested in researching this phenomenon and has undertaken many studies toward understanding this hazard and finding methods by which it can be combated. One of the tools employed by the Space Agency is surface electromyography (SEMG), a noninvasive method for analyzing and recording the conditions of muscles while at rest and in use. SEMG uses, as the name implies, surface electrodes to monitor and graph muscle activity.

Partnership

NASA funded a study in which researchers from Boston University and the Massachusetts Institute of Technology used SEMG to measure the muscle activity of astronauts during strenuous activities directly before and after space shuttle missions. The research, done at the NeuroMuscular Research Center at Boston University under the direction of engineer Lee Brody, resulted in the design and implementation of two experimental protocols. To determine and map changes in muscles of shuttle astronauts, the team designed a protocol to measure signals from a leg muscle during a series of exercises.

For the other project, the team designed a measurement technique and accompanying software to test different designs for the gloves astronauts would wear during space walks. The gloves astronauts were using previously led to extreme hand and forearm fatigue while they conducted repairs and other tasks that required precision hand use. The newer gloves alleviate these conditions.

Since completing the NASA-funded research projects, Brody has found additional applications for the methods and techniques he learned during the experiments. He is now the chief engineer for the Chiropractic Leadership Alliance (CLA), of Mahwah, New Jersey, a professional organization serving the chiropractic community.

Product Outcome

Founded by Drs. Patrick Gentempo and Christopher Kent, the CLA is dedicated to advancing the study and practice of chiropractic medicine. Its space experiment-based technology is called the Insight Subluxation Station and is a neurospinal screening and evaluation system based on the NASA research conducted by Brody and the SEMG methodology. While many chiropractic groups use SEMG, this is the only system employing the unique NASA technology. In fact, the CLA has even secured exclusive certification for the product with the Space Foundation, a nonprofit organization that promotes awareness and advancement of space-related research and exploration. Even more, the technology, available worldwide, is the only chiropractic device of its nature whose applications, protocols, indications, and normative data have all been published in peer-reviewed literature and are taught at certified chiropractic universities and colleges.

In the human body, the central nervous system is composed of three parts: the brain, the spinal cord, and the nerves. Stress (whether physical, psychological, or physiological) in any of these areas can affect the others and throw the whole system off, often resulting in back pain. The role of the chiropractor is to repair disruptions to the spine and explain to patients the connections between

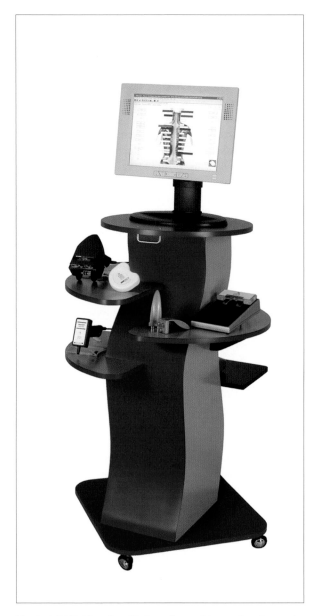

The Insight Subluxation Station allows chiropractors to create visual representations of abnormal nerve functions.

the parts of nerves and back pain. Toward this end, chiropractors employ a wide variety of diagnostic tests and techniques to examine the spine and nervous system, but finding clear evidence of nerve disruption can, nonetheless, prove difficult.

Chiropractors use their hands, checking the muscles and spinal alignment, and they can take X-rays of vertebrae and analyze spinal degeneration, but it is still difficult to quantify the amount of nerve disruption. Just as important, it is also difficult for chiropractors to check the results of various types of treatment. The Insight Subluxation Station addresses that shortfall.

The Insight Subluxation Station employs a scanning device that measures differences in skin temperature and the sensitivity of paraspinal tissue and assesses range of motion, variations in heart rate, and the amount of tension of paraspinal muscle activity. It then, through unique software, creates visual representations of abnormal nerve functions.

These visual representations allow chiropractors, for the first time, to be able to actually see disruptions in nerve functioning. The color-coded graphs allow chiropractors to determine what methods of care will be best to pursue, and also allow caregivers to follow-up by assessing the effectiveness of the chosen treatment. Furthermore, the computer-generated charts show patients exactly where the nerve dysfunction exists, which helps the patient understand the reasoning behind different treatment methods.

A chart also helps with patient retention. Since the report usually validates the patient's concerns, he or she is more likely to feel assured that the treatment is based on something other than manual poking and prodding, and upon return, knows that a visual report of progress will be received.

The system, which has settings that allow it to be used on infants, children, and adults, is readily compatible with most chiropractic offices and cleared by the U.S. Food and Drug Administration. It uses a standard electrical power source and features USB communication technology to allow it to interface quickly and effectively with other devices. The software is compliant with Federal standards set by the Health Insurance Portability and Accountability Act (HIPAA), an initiative to regulate and standardize electronic health care information exchange.

The CLA has created a 4-day training program for chiropractors who purchase the Insight scanning device. What Gentempo refers to as a sort of "boot camp" for chiropractors, the training session, called the Total Solutions package, teaches chiropractors not only how to use the device, but how to get the most out of their chiropractic practices. ❖

Insight Subluxation Station™ and Total Solution™ are trademarks of the Chiropractic Leadership Alliance.

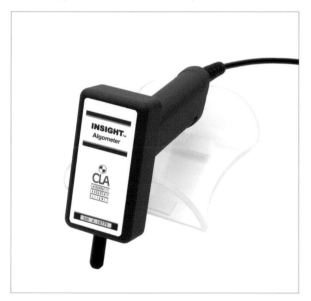

The Insight Algometer (above) is a hand-held device that accurately measures the sensitivity of paraspinal tissues. Tender areas discovered by palpation can be quantified, and progress can be graphically displayed throughout a course of care.

The Insight Inclinometer (left) performs a standard two-point range-of-motion (ROM) assessment. The inclinometry protocol evaluates end-point ROM as compared to normative data, which then provides the percentage of impairment. The Pulse Wave Profiler (above) measures heart rate variability, the beat-to-beat variation in heart rate. This natural rise and fall of heart rate is caused by several physiologic phenomena, including breathing and autonomic nervous system activity. Chiropractors use this measurement to get a window into how the autonomic nervous system modulates heart rate in the baseline or resting state.

Hand-Held Instrument Fights Acne, Tops Over-the-Counter Market

Originating Technology/NASA Contribution

When the U.S. Congress created NASA in 1958, it sought to ensure that the Nation saw returns on its investments in aerospace research. It, therefore, wrote provisions into the Space Act Agreement that formed the Agency requiring that NASA share its technological advances and engineering expertise with the American public. Since that time, the NASA scientific prowess that sends people and equipment up into space has also come back down to Earth, in the form of products and processes that make everyday life better all around the globe.

Zeno is a hand-held, portable electronic medical device that is clinically proven to make pimples disappear fast. In fact, for treating acne pimples, it is the most scientifically advanced and effective device available without a prescription. It applies a precisely controlled heat dose directly to the pimple through a metal pad. One treatment lasts 2.5 minutes.

Through a variety of methods, industry can work with the Space Agency to access the cutting-edge resources and advanced technologies created as a result of the Nation's investment in space. For companies that do not need NASA-developed technology or advanced test facilities but still want to leverage some of the brain power of NASA's engineering teams to solve a design problem in their product, NASA funds the Space Alliance Technology Outreach Program, more commonly referred to as "SATOP."

SATOP, administered by Bay Area Houston Economic Partnership, is a cooperative program between Florida, New Mexico, New York, and Texas. It brings together more than 55 space companies, universities, colleges, and NASA centers to make the expertise garnered through the Nation's investment in space exploration available to small businesses. SATOP finds professionals within these organizations who volunteer their time and expertise in solving various design and engineering challenges.

Any type of small business is encouraged to submit a technical challenge to SATOP. If SATOP is able to assist, the small business is provided with up to 40 hours of free technical assistance from a scientist or engineer in the Space Program.

Partnership

Tyrell Inc., a Houston-based medical technologies company, was able to access engineering support for a problem with redesigning a heating element for a hand-held acne-fighting device. The device was born of necessity, when Tyrell founder Robert Conrad, now the company's chief operating officer, was working at a biological testing firm. One of the experiments he was conducting involved working with proteins, which included growing bacterial colonies, shocking them with heat to kill the bacteria, and then extracting the proteins. Conrad, who had been plagued with adult acne and was looking for a solution, recalled while conducting these experiments that acne itself is caused by the bacteria, *P. acnes*, and wondered about the possibility of shocking pimples with heat.

At home in his garage, he began experimenting with time and temperature curves to develop a device that would shock and kill acne-causing bacteria. He built a working prototype and was content to stop there and keep the device for personal use. Family and friends, however, took interest in the device, and each wanted one. Conrad saw the obvious market potential and the need that people, like him, had for this device.

While the prototype proved feasible, the design still needed some tweaking; the device was cumbersome and expensive to produce. Conrad was introduced to SATOP through the Houston Technology Center, a local technology business accelerator with a Space Act Agreement with NASA's Johnson Space Center.

Conrad presented the acne-clearing device and explained that he wanted a new heating element design that would be resistant to oils and acids, with a cylindrical shape that would be 0.0312 inch in diameter and would provide temperature feedback through resistance change. In addition, the heater needed to run on no more than 5 volts, consume approximately 5 watts during operation, and, most importantly, be cost competitive.

SATOP directed Conrad to Allen J. Saad, of The Boeing Company, a principal design engineer at Kennedy Space Center. Saad assessed the design and made several major contributions. He was able to give the company engineering advice toward designing a new element that reduced the cost of the heating element from about $80 each to just 10 cents, making the product more marketable. Shortly after meeting with Saad, Conrad was able to implement the design modifications and take the product to market. It is now the highest selling over-the-counter medical device for the treatment of acne.

In 2006, the product, dubbed Zeno, was named the "SATOP Texas Success Story of the Year." Conrad had positive things to say about the partnership as well: "Without the help I received from SATOP, the product would likely still be in my garage. Instead, we now have three offices, and sales are skyrocketing."

The company has expanded greatly, with offices in Houston and Seattle, and a manufacturing facility in Penang, Malaysia, that country's only U.S. Food and Drug Administration (FDA)-approved facility.

Product Outcome

According to Tyrell, nearly 90 million people in the United States spend more than $2 billion a year on acne treatments, with limited results and sometimes significant side effects. Its over-the-counter device, Zeno, has been cleared by the FDA, and is on the market to offer an effective solution to this widespread problem.

Zeno employs proprietary ClearPoint technology to provide relief of mild to moderate inflammatory acne. It delivers a precisely controlled low-level dosage of heat to the blemish, causing the bacteria at the root of more than 90 percent of acne to self-destruct. The heat is the exact temperature needed to kill the bacteria while still being below the point at which healthy skin would be damaged, and an internal microprocessor continually modulates the temperature to match the heat absorption of the individual user. Replaceable treatment tips ensure thermal efficiency over the life of the product.

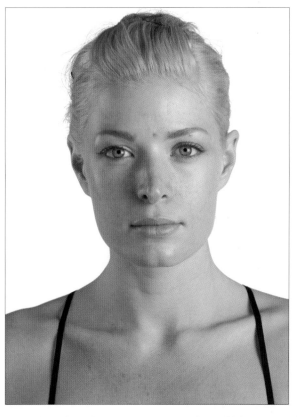

Two to three 2.5-minute treatments spread over 24 hours are sufficient for most pimples. Often, only one treatment is needed.

Zeno spot-treats acne. Clinical trials showed that 90 percent of blemishes treated with Zeno faded or disappeared within 24 to 48 hours. Over the course of 12 to 24 hours, the user applies the device to the affected area for two or three treatment cycles of 2.5 minutes each. That is usually all that it takes, though it can also still be safely used in conjunction with other acne treatments, including over-the-counter and prescription medications.

Within its first year on the market, Zeno was cited by various publications for several awards, including Allure's 2005 "Best of Beauty," Marie Claire's "10 Best Gadgets for Girls," and Popular Science's 2005 "Best of What's New." It is available in three models: Zeno, Zeno Pro, and Zeno MD.

A variation of the Zeno for use in treating herpetic lesions such as cold sores, by killing the virus that causes them, is currently undergoing FDA trials. ❖

Zeno® and ClearPoint™ are trademarks of Tyrell Inc.

Multispectral Imaging Broadens Cellular Analysis

Originating Technology/NASA Contribution

How does life begin and evolve? Does life exist elsewhere in the universe? What is the future of life on Earth and beyond?

These questions have lingered for many years, and we still do not fully know the answers to them. NASA, however, in its efforts to sustain life here on our home planet and chart a course for humans to explore the Moon, Mars, and beyond, is going to new depths to better address them—deep beneath the ocean surface.

In the floors of our oceans are holes that spew hot, gaseous, mineral-rich liquids from the deep, subsurface magma below. Scientists from NASA's Advanced Life Support Group at Ames Research Center are hoping that these holes, called hydrothermal vents, may not only help them unlock the secrets of the deep, but help them learn if exotic life forms exist on other planetary bodies, including Mars and Jupiter's moon Europa.

Because hydrothermal vents are thousands of meters underwater, they are not exposed to sunlight. Without sunlight, there may be abiotic life forms—life forms that exist devoid of photosynthetic input or decomposition of organic materials—in or near these vents. NASA scientists hypothesize that the deep subsurface of Earth could be home to organisms that exist solely on chemical energy that is generated from the off-gassing magma below the ocean floor. They further hope that this extreme underground environment could provide insight into whether similar life forms exist elsewhere in the universe, in environments that are far removed from the Sun and therefore also require other sources of energy. For example, scientists have targeted Europa because of its thick ice crusts and the mounting evidence that points to there being an ice-covered ocean that could potentially harbor similar hydrothermal vents.

To test their hypothesis, the Advanced Life Support Group scientists have built a life-detection instrument called Medusa to collect, store, and analyze sample

It is possible that beneath Europa's surface ice, there is a layer of water, kept liquid by tidally generated heat. If so, it would be the only place in the solar system besides Earth where liquid water exists in significant quantities.

organisms from erupting hydrothermal vents. For the sample analyses, Medusa is equipped with a spectral-analysis chemical sensor that uses a process called flow cytometry to examine the natural glow of light, or fluorescence, emitted from any of the samples collected as ocean water flows through the instrument. When the scientists retrieve the samples, they inject a dye into them that interacts with the fluorescence and emits colors to reveal their chemical composition.

Medusa has already been deployed several times to study hydrothermal vents, and scientists will continue to send the instrument to the bottom of the ocean in an effort to validate their theory. NASA also plans to test the instrument in other extreme environments of Earth in its continuing quest to seek unknown life forms. Meanwhile, scientists are working to apply the spectral-analysis capabilities of Medusa's chemical sensor to other areas of research, especially in studying the effects of gravity and cosmic radiation on human cells to possibly create an enhanced system for monitoring the biological effects of space travel on astronauts.

Partnership

To develop the advanced flow cytometry process that is at the crux of Medusa and may one day be part of an advanced astronaut health-monitoring system, NASA sought the help of private industry. The Agency issued a solicitation through the **Small Business Innovation Research (SBIR)** program at Ames to find a partner for the project. Specifically, NASA needed a partner that could produce a high-speed flow cytometry process to continuously monitor cells for anomalies and bacteria growth. The system would ideally be able to image cells with high-fluorescence sensitivity and would be tolerant of wide variations in sample concentration and other characteristics by having a wide depth of field, allowing cells to be kept in focus regardless of where they happened to be in the flow stream.

Flow cytometry is a powerful technique, but it has several limitations that hinder its use for NASA projects. Most commercial flow cytometers cannot image cells like a microscope. Instead, they measure only the total amount

NASA's Medusa instrument inside the submergence vehicle Alvin's instrumentation basket in preparation for launch.

The submergence vehicle Alvin as it is about to dive to the bottom of the ocean. Released from the mothership, the Research Vessel Atlantis (operated by Woods Hole Oceanographic Institution, of Woods Hole, Massachusetts) with the diver standing on top.

of fluorescence emitted by each cell. Because they do not have the ability to determine where in the cell the signal is coming from, their applications in cell biology are mainly limited to measuring the total quantities of specific molecules in or on the cell. A few flow cytometry systems can produce images of cells in flow using transmitted light, but these systems lack the sensitivity necessary to image faint fluorescence.

Flow cytometers also generally require that cells flow through the center of a tightly focused laser beam, which can make them vulnerable to misalignment. Designs that are less sensitive to misalignment tend to sacrifice fluorescence sensitivity.

Many flow cytometry systems are also just too big and impractical for NASA's purposes, especially since they incorporate pressurized fluid vessels that employ gravity and high pressure to drive the sample through the system.

Amnis Corporation, a Seattle-based biotechnology company, developed a technology called ImageStream for producing sensitive fluorescence images of cells in flow, and happened to be seeking ways to get whole cells into focus in order to increase the usefulness of its systems for research applications. The company had several ideas for how to achieve an extended depth of field, all of which required a level of funding that was just not in its budget. When Amnis heard about the SBIR solicitation, however, the realization came that it could be the perfect opportunity to reap the funding necessary to develop extended depth-of-field technology. The company responded to the SBIR solicitation and proposed to evaluate several methods of extending the depth of field for its ImageStream system, pick the best method, and implement it as an upgrade to its commercial products. This would allow users to view whole cells at the same time, rather than just one section of each cell.

Through Phase I and II SBIR contracts, Ames provided Amnis the funding the company needed to develop this extended functionality. For NASA, the resulting high-speed image flow cytometry process made its way into Medusa and has the potential to benefit space flight health monitoring. On the commercial end, Amnis has implemented the process into its flagship product, ImageStream.

Product Outcome

ImageStream combines high-resolution microscopy and flow cytometry in a single instrument, giving researchers the power to conduct quantitative analyses of individual cells and cell populations at the same time, in the same experiment. Incorporating the extended depth of field developed during the SBIR project, ImageStream is designed to provide multispectral images of rapidly moving objects (in flow) with very high sensitivity, producing up to 6 different microscopic images of each cell flowing through the instrument, at a rate of 15,000

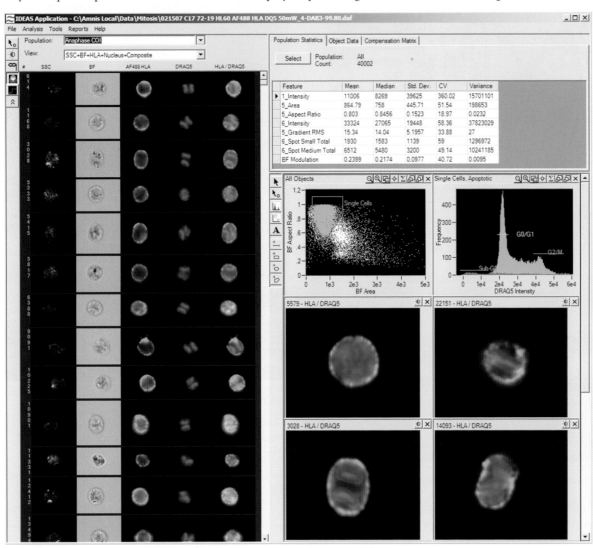

Screen shot of Amnis Corporation's IDEAS data analysis software showing the detection of cells in division.

cells per minute. It captures the images at high speeds with a built-in, charge coupled device camera that electronically tracks the motion of cells. A sophisticated auto-focus system continually optimizes image quality.

Getting a whole cell into focus is vital for many different types of cellular analyses, especially for a technique known as fluorescence in situ hybridization (FISH). FISH involves the binding of engineered genetic probes to specific genes or other DNA sequences within cells. The genetic probes consist of a short piece of DNA attached to a fluorescent molecule. The DNA part finds and binds to its complementary DNA sequence(s) within the cell's genome and the fluorescent molecule signals that the binding has taken place, producing a small spot of light within the cell. By detecting and counting these "FISH" spots, a clinician or a researcher conducting an analysis can tell, for instance, if a patient has extra copies of a gene or a chromosome, whether there is damage to their genes, or whether they have lost genetic material—all of which are very useful for detecting cancer and birth defects.

FISH analysis is primarily a manual process. Clinicians and researchers have to stare through microscopes, adjusting the focus knob until they find all the FISH spots within a given cell. When they do find the spots, they have to manually count them (for about 50 cells per diagnostic test, on average). According to Amnis, this methodology has several problems: the test is slow and expensive, the counting is subjective and prone to error, and, unless a genetic abnormality is present in at least 10 percent of the cells analyzed, researchers will not likely catch it. Automated microscopy systems have been applied to the problem, but they are forced to take multiple images of each cell at different focus settings to ensure that all FISH spots are detected, slowing the test and reducing the economic advantage of automation, according to Amnis.

Prior to the incorporation of the extended depth-of-field technology into the ImageStream system, it was common for one or more FISH spots within a cell to be out of focus as the cell passed through the system.

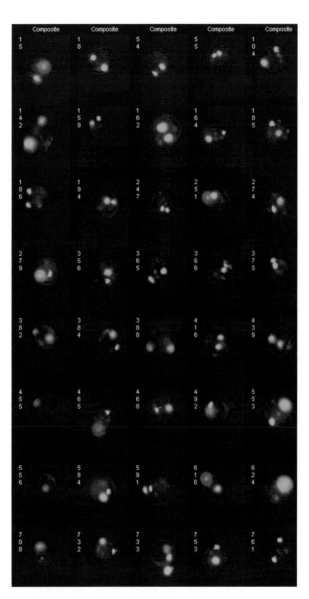

Extended depth-of-field (EDF) image of FISH probes dispersed in the nucleus along the optic axis or in close proximity in the lateral direction. EDF improves image presentation by removing focus-induced variation.

By enhancing the depth of field, Amnis has upgraded ImageStream to accurately detect and count FISH spots through an automated process that eliminates the subjectivity associated with manual counting and can be performed on thousands of cells per minute. This helps drive down the cost of testing by eliminating manual labor, reducing the time needed to perform the test, and increasing accuracy to eliminate retests and false results. Amnis recently announced the availability of this new, high-throughput process.

ImageStream is also built for many other applications, including cell signaling and pathway analysis; classification and characterization of peripheral blood mononuclear cell populations; quantitative morphology; apoptosis (cell death) assays; gene expression analysis; analysis of cell conjugates; molecular distribution; and receptor mapping and distribution. These applications are practical for advanced research in the fields of hematology, immunology, and oncology.

As an add-on option, ImageStream users can pair the instrument with Amnis' statistical image-analysis software package, IDEAS. This statistical analysis tool is extremely robust, as it provides more than 200 features for every cell analyzed. These features can be used by clinicians and researchers to generate histograms and scatter plots for graphical identification and representation of cells and cell populations based on characteristics like fluorescence intensity, size, shape, texture, probe distribution heterogeneity, and co-localization of multiple probes. ❖

ImageStream® and IDEAS® are registered trademarks of Amnis Corporation.

Hierarchical Segmentation Enhances Diagnostic Imaging

Originating Technology/NASA Contribution

Developed by Dr. James Tilton, a computer engineer with Goddard Space Flight Center's Computational and Information Sciences and Technology Office, Hierarchical Segmentation (HSEG) software allows for advanced image analysis. The software organizes an image's pixels into regions based on their spectral similarity, so rather than focusing on individual pixels, HSEG focuses on image regions—and how they change from a coarse-to-fine perspective. These regions (segmentations) are at several levels of detail (hierarchies) in which the coarser segmentations can be produced from the finer-resolution segmentations by selective merging of regions. In addition, the segmentation hierarchies provide analysis clues through the behavior of the image region characteristics over several levels of segmentation detail. Thus, by enabling region-based analysis, the segmentation hierarchies organize image data in a manner that makes an image's information content more accessible.

Within NASA, HSEG was used to identify and extract magnetospheric radio-echo and natural plasma-wave signals captured by the Radio Plasma Imager aboard the Imager for Magnetopause-to-Aurora Global Exploration (IMAGE) spacecraft, which was launched in March 2000, and after completing its 2-year primary mission, was lost in space after a full 5.8 years of exploration.

To advance the usability of the HSEG, Tilton also developed Recursive Hierarchical Segmenting (RHSEG) Pre-processing Software, which significantly improves the extraction of patterns from complex data sets. Optimized for speed and accuracy, the patent-pending algorithm that fuels the software provides the user with precise control for selecting the desired level of detail from the hierarchy of results. The software allows the user to group non-spatially adjacent regions for flexibility with a wide range of image and data types, providing accurate graphic representation of imagery data with minimal distortion and a fine resolution of detail.

Although originally developed for NASA's remote-sensing applications, HSEG has proven its value in other realms as well. Also originally designed for remote Earth sensing, the RHSEG Pre-processing Software is broadly applicable for a variety of uses, from medical imaging to data mining, and it is now capable of three-dimensional (3-D) data analysis, a capability NASA is in the process of patenting.

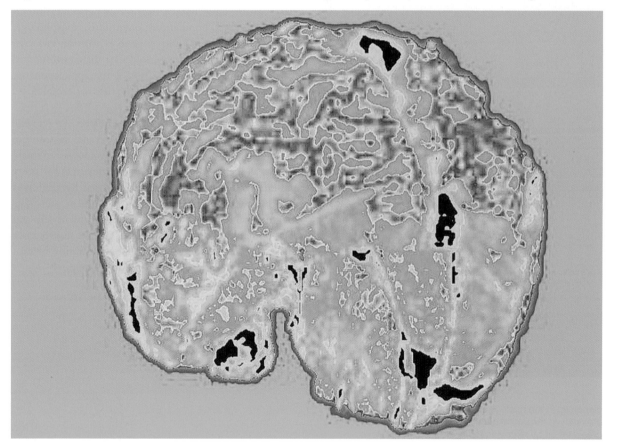

Rather than working on the usual pixel-by-pixel basis, the RHSEG automatically organizes pixels into regions hierarchically, based on their spectral similarity. Looking at these regions, as opposed to individual pixels, allows the user to isolate specific features that are impossible to distinguish by other methods. Thus, the RHSEG provides a more reliable and accurate understanding of the image.

Partnership

Bartron Medical Imaging LLC (BMI), of New Haven, Connecticut, gained a nonexclusive license from Goddard Space Flight Center to use the RHSEG software in medical imaging.

In order to manage the image data, BMI also licensed two pattern-matching software programs from NASA's Jet Propulsion Laboratory that were used in image analysis and three data-mining and edge-detection programs from Kennedy Space Center. The Kennedy imaging technologies are currently in use at NASA to identify and track foreign object debris during space shuttle liftoff, and in the Cable and Line Inspection Mechanism used to test the shuttle's emergency escape system.

In addition, BMI made NASA history by being the first company to partner with the Space Agency through a Cooperative Research and Development Agreement (CRADA). This agreement, for development of the 3-D version of RHSEG, grants (in advance) a partially exclusive license for the resulting technology patents within BMI's fields of use (i.e., the diagnosis and treatment of breast cancer, cervical cancer, brain cancer, heart disease, osteoporosis, and periodontal diseases).

Product Outcome

With U.S. Food and Drug Administration clearance, BMI will sell its Med-Seg imaging system with the 2-D version of the RHSEG software, licensed from NASA in 2002 and featured in *Spinoff* 2004. The device is intended to analyze medical imagery from computed tomography (CT or CAT) scans, positron emission tomography (PET) scans, magnetic resonance imaging (MRI), ultrasound, digitized X-rays, digitized mammographies, dental X-rays, soft tissue analyses, and moving object analyses; the technology is also equipped to evaluate soft-tissue slides such as Pap smears for the diagnoses and management of diseases. The advanced image segmentations produced by the RHSEG software allow the Med-Seg system to bring

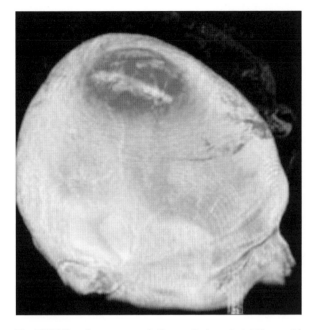 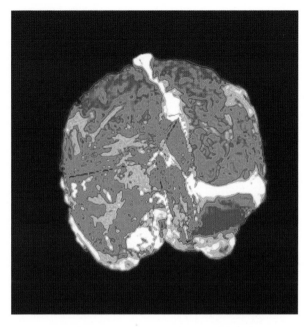

The RHSEG software presents its results in a straightforward format consisting of a hierarchical set of image segmentations in either two or three spatial dimensions. This hierarchical presentation of results allows the user to choose the segmentation(s) of interest and to perform additional analyses.

out details in these tests not previously seen with the naked eye. This allows for quick and accurate diagnosis of diseases. Additionally, unlike some other image-analysis devices, Med-Seg does not manipulate the image, so what the physician sees is truly what is there, providing truer images than many other imaging techniques.

The Med-Seg system reveals image properties not seen with even up-to-date image enhancement techniques. It is available with a custom data-mining server; a computational engine; and an object-oriented database to manage data, models, and results, and access BMI's Hierarchical Image Segmentation Beowulf Cluster, an online problem-solving network.

By extending the software's capabilities to three dimensions, BMI's device may be able to produce a pixel-level view of all sides of a tumor or lesion. While current technology can produce 3-D imagery, the RHSEG software will be able to segment an image in ways that more clearly define problem areas. For example, the 3-D version of Med-Seg may be able to identify very early buildup of soft plaque within the arteries or identify density levels of microcalcification in mammographies, allowing physicians to diagnose malignant breast tumors well before they would normally be seen. In brain images, the physicians using Med-Seg will also be able to make earlier diagnoses of tumors or arteriovenous malformations. ❖

Med-Seg™ is a trademark of Bartron Medical Imaging LLC.

Transportation

NASA technologies enhance ways that we travel. The benefits featured in this section:

- Ease air traffic management
- Advance rotorcraft design
- Improve flight safety
- Boost helicopter performance
- Protect general aviation aircraft

Comprehensive Software Eases Air Traffic Management

Originating Technology/NASA Contribution

Gridlock, bottlenecks, bumper-to-bumper jams—we all get caught in congestion at one time or another, as the rigors of road traffic are an inevitable part of life. Sometimes we do our best to get ahead, taking advantage of the slightest opening in the next lane, in anticipation that it is moving quicker than the snail's pace of our current position. Other times, we just patiently ride it out, opting to sit back and get comfortable, fully surrendering to the sea of cars and trucks ahead.

And then there are the times that, in all desperation, we look up into the sky, see all of that available space, and ask ourselves when the future is going to bestow upon us the fantastic gift of flying cars that would help open things up and let us bypass all of the congestion on the ground.

Airspace, however, is not much different than road space, in terms of congestion. As much as we like to think that there is clear, blue sky as far as the eye can see, the truth of the matter is that, with thousands of planes flying overhead in the United States at any given time, there can sometimes be just as many traffic and delay problems associated with air travel as there can be with ground travel.

To keep a handle on the complex flow of aircraft, the United States depends on a tightly run air traffic control system. Air traffic control centers around the country work to help aircraft maintain safe distances while in flight, as well as during takeoffs and landings, to prevent accidents. In addition, air traffic control centers work to keep in-flight pilots informed of changing weather conditions that may impact their flight paths. Essentially, the air traffic control centers' main objectives are to maximize safety and minimize delays in the air and at U.S. airports and airfields.

To help air traffic control centers improve the safety and the efficiency of the National Airspace System (the term used for the overall environment in which aircraft operate throughout the United States), Ames Research Center developed software called the Future Air Traffic Management Concepts Evaluation Tool (FACET). With powerful modeling and simulation capabilities, FACET can swiftly generate thousands of aircraft trajectories (as many as 15,000 on a single computer) that can help to streamline the flow of air traffic across the entire National Airspace System. Actual air traffic data and weather information are utilized to evaluate an aircraft's flight-plan route and predict its trajectories for the climb, cruise, and descent phases. The dynamics for heading (the direction the aircraft nose is pointing) and airspeed are also modeled by the FACET software, while performance parameters such as climb/descent rates and speeds and cruise speeds can also be obtained from data tables. The resulting trajectories and traffic flow data are presented in a 3-D graphical user interface.

FACET is one of the many air traffic management software tools developed at Ames as part of NASA's Airspace Systems Program, which aims to satisfy the Nation's plans for a next-generation airspace system. It has the distinction, however, of being the winning software in NASA's 2006 "Software of the Year" competition.

Partnership

In 2005, Ames licensed FACET to Flight Explorer Inc., for integration with its Flight Explorer version 6.0 software system. According to the McLean, Virginia-based company, it is the world's leading provider of real-time global flight tracking information, reporting, and display products. Its clients include over 80 percent of major North American airlines and 22 of the top 30 regional airlines. It also provides a host of free services, including online flight tracking, airport information tools, and daily air travel reports.

The primary FACET features incorporated in the Flight Explorer software system alert airspace users to forecasted demand and capacity imbalances. By having advanced access to this information, dispatchers can anticipate congested sectors (airspace) and delays at airports and decide if they need to reroute flights. Overall, the FACET developers at Ames assert that airspace users can use this information to develop enhanced flight-routing strategies that save fuel, preserve airline schedules, and reduce passenger delays and missed connections.

Product Outcome

FACET is now a fully integrated feature in the Flight Explorer Professional Edition (version 7.0). Flight Explorer Professional is a flight-tracking and management-decision support tool that the aviation community can use to improve operational efficiency and business performance. It incorporates NASA's FACET technology to graphically depict airports and air sectors that are approaching capacity or are over capacity. FACET provides a count of the total number of arrivals and departures at airports every 15 minutes, plus a count of aircraft flying and aircraft anticipated to be flying within a given sector every 15 minutes—all while calculating loading predictions and weather conditions—to help keep flights on schedule. Data is derived from posted flight plans, and information pertaining to airports and sectors is color-coded, based on capacity. For instance, if an airport is at less than 80-percent capacity, it is marked green; if it is at greater than or equal to 80-percent capacity, it is marked yellow; if it is at 100-percent capacity, it is marked red. Green, yellow, and red color codes also apply for the capacity of sectors.

Flight Explorer Professional offers end users a plethora of other benefits, including ease of operation. As a computer-based, graphical aircraft situation display, it uses a standard Internet connection to securely stream real-time aircraft and weather information from the Flight Explorer Inc. data center to the user, every 10 seconds. The information the data center collects is retrieved from radar, satellite, and other tracking mechanisms provided by the Federal Aviation Administration and other aviation sources.

The software's Flight Alert System generates automatic alerts to inform users of important events, better preparing them for weather conditions and potential airport delays. A real-time log of any significant events occurring during flight or at an airport is recorded and can be exported to a database for analysis. For the weather alerts, Flight Explorer depends on the NEXRAD (next generation radar) weather-surveillance satellite to display forecast scenarios 6 hours ahead of time. Graphical overlays are also built in so that users can view real-time maps containing layers of information on top of each other, such as air traffic patterns on top of regional weather conditions.

When asked about the vision for the Flight Explorer software, Jim Kelly, Flight Explorer Inc.'s chief executive officer, said, "It's pretty simple. Start with the aircraft situation display as a cornerstone for display and messaging services, make sure it is the best out there by listening to customers, and partner with those willing to help adapt and improve it."

Evidently, the customers agree: "Nothing compares to the comprehensive feature set and ease of use of Flight Explorer Professional," noted Casey Barr, owner services manager for Regal Aviation, a private jet service based in Dallas.

"We utilize Flight Explorer Professional in both our daily operation and analysis. It allows our operational managers to understand and then react to situations in real time," stated Christopher Forshier, of the Systems Operations Coordination Center at Houston-based Continental Airlines.

Flight Explorer Professional also provides international, real-time flight coverage over Canada, the United Kingdom, New Zealand, and sections of the Atlantic and Pacific Oceans. In addition, Flight Explorer Inc. has broadened coverage by partnering with Honeywell International Inc.'s Global Data Center, Blue Sky Network, Sky Connect LLC, SITA, ARINC Incorporated, Latitude Technologies Corporation, and Wingspeed Corporation, to track their aircraft anywhere in the world. ❖

Flight Explorer® and Flight Explorer Professional® are registered trademarks, and Flight Alert™ is a trademark of Flight Explorer Inc.

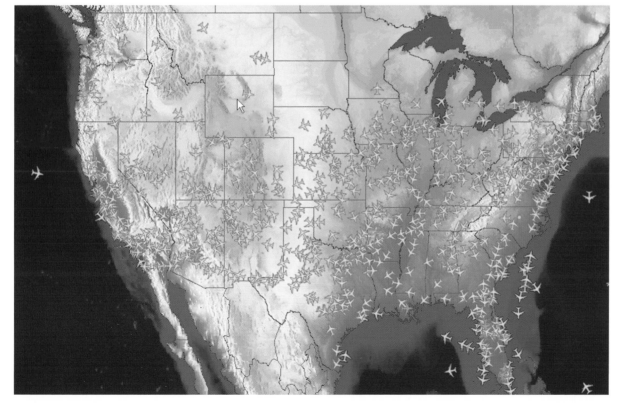

The Future Air Traffic Management Concepts Evaluation Tool (FACET), developed at Ames Research Center, was selected as NASA's 2006 "Software of the Year." FACET is a flexible software tool that provides powerful simulation capabilities and can rapidly generate thousands of aircraft trajectories to enable efficient planning of traffic flows at the national level.

Modeling Tool Advances Rotorcraft Design

Originating Technology/NASA Contribution

Often times, when people think of NASA, they think of space travel. The first "A" in NASA, however, is for "Aeronautics," and the Agency has always held as one of its tenets to explore, define, and solve issues in aircraft design. Just as often as NASA is associated with space travel, when people hear aeronautics, they often think of airplanes, but part of NASA's aeronautics program is one of the most advanced rotorcraft design and test programs in the world.

Located at Ames Research Center, the Aeromechanics Branch of the Flight Vehicle Research and Technology Division conducts theoretical and experimental research in support of the U.S. helicopter industry and the U.S. Department of Defense. At this research site, engineers study all aspects of the rotorcraft that directly influence the vehicle's performance, structural and dynamic responses, external acoustics, vibration, and aeroelastic stability.

They use modern wind tunnels and advanced computational methodologies to calculate fluid dynamics and perform multidisciplinary, comprehensive analyses in the quest to further understand the complete rotorcraft's operating environment and to develop analytical models to predict aerodynamic, aeroacoustic, and dynamic behavior. The experimental research also seeks to obtain accurate data to validate these analyses, investigate phenomena currently beyond predictive capability, and achieve rapid solutions to flight vehicle problems.

Partnership

Founded in 1979, Continuum Dynamics Inc. (CDI), of Ewing, New Jersey, specializes in advanced engineering services, including fluid dynamic modeling and analysis for aeronautics research. Its clients include government agencies, as well as the aerospace, nuclear, and pharmaceutical industries, and it has been partnering with NASA since its inception.

The company has converted years of NASA-funded research efforts into a variety of commercial products. For example, 1987 and 1992 NASA **Small Business Innovation Research (SBIR)** grants on helicopter wake modeling resulted in software code used in a blade redesign program for Carson Helicopters, of Perkasie, Pennsylvania, that simultaneously increased the payload of its Sikorsky S-61 helicopter by 2,000 pounds and increased cruise speeds at 10,000 feet by 15 knots.

Follow-on development of this same rotorcraft model, based on 1999 and 2002 NASA SBIR work, resulted in a $24 million revenue increase for Sikorsky Aircraft Corporation, of Stratford, Connecticut, as part of the company's rotor design efforts.

Altogether, the company has completed a number of SBIR projects with NASA, including early rotorcraft work done through Langley Research Center, but more recently, out of Ames.

This rotorcraft model software code, marketed by CDI as the Comprehensive Hierarchical Aeromechanics Rotorcraft Model (CHARM), is a tool for studying helicopter and tiltrotor unsteady free wake modeling, including distributed and integrated loads, and performance prediction.

Product Outcome

Under continuous development at CDI for more than 25 years, CHARM analyzes the complete aerodynamics and dynamics of rotorcraft in general flight conditions. CHARM has been used to model a broad spectrum of rotorcraft attributes, including performance, blade loading, blade-vortex interaction noise, air flow fields, and hub loads. The highly accurate software is currently in use by all major rotorcraft manufacturers, NASA, the U.S. Army, and the U.S. Navy.

Available as a stand-alone product or adaptable to existing simulator and analysis systems, this software code is well suited for performing analysis on advanced aerodynamic design as well as for research on new designs.

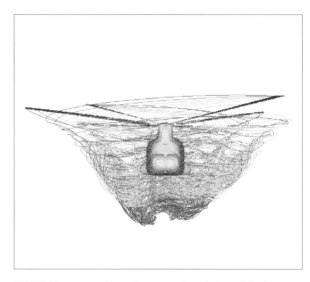

CHARM is a computer software product that models the complete aerodynamics and dynamics of rotorcraft in general flight conditions.

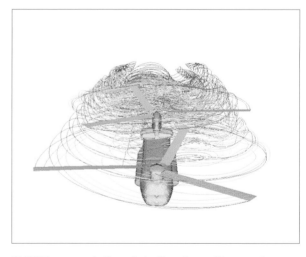

CHARM represents the culmination of over 25 years of continuous development of rotorcraft modeling technologies at Continuum Dynamics Inc., and incorporates landmark technical achievements from a variety of NASA, U.S. Department of Defense, and company-sponsored initiatives.

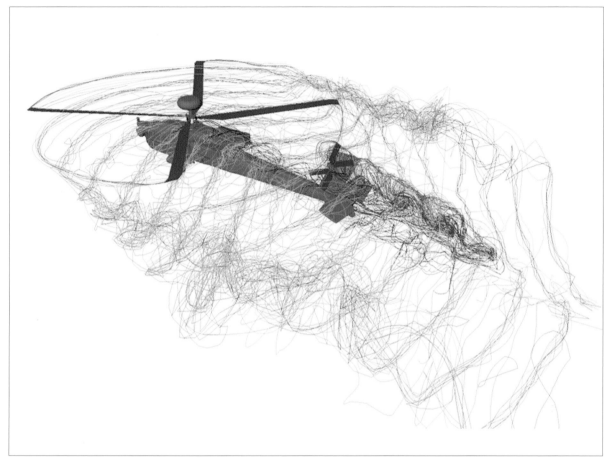

CHARM is available for licensing as a stand-alone analysis program or as a module that couples directly into existing flight simulation or analysis software.

This increased performance allows the designer to explore high-risk technologies, expand design parameter ranges, and evaluate critical components at a level of detail never before possible.

According to CDI, CHARM features the most advanced free vortex wake model currently available that directly computes wake roll-up and vortex core properties from first principles. In addition, no empirical constants are required, which is a distinct advantage for design work and modeling transient maneuvering flight for which wake characteristics are continually changing. The software code has also been coupled with acoustic prediction software to provide fast prediction of rotor loading and thickness noise.

CHARM's lifting surface blade aerodynamics model is well suited for analysis of complex tip shapes and other 3-D effects, and its coupled wake/panel calculation is unique—incorporating state-of-the-art fast vortex and fast panel methods to allow fully coupled rotor/wake/airframe solutions.

Most recently, the software has been incorporated into a joint CDI-Army study for incorporation into flight simulation software to practice landing rotorcraft in sandy or dusty conditions. Known as brownout, the effect of sand, dust, and debris kicked up by a helicopter during takeoff and landing can temporarily blind a pilot, as well as damage moving parts. The simulator is designed to identify, characterize, and evaluate brownout conditions for general aircraft, wind, ground topology, and flight maneuvers. The study seeks to improve pilot training for these dangerous conditions, help in mission planning, decrease damage to rotorcraft components, and improve sensor technologies most affected by these conditions. ❖

The software includes extensive 3-D graphics capabilities that allow highly detailed visualization of the rotor/wake motion and wake/surface interaction. Easy to use and reliable, CHARM plugs directly into alternate rotorcraft analysis or flight simulation software.

CHARM incorporates one of the most advanced wake models currently available, combining a full-span, freely distorting, constant vorticity contour wake model and an analytical tip vortex roll-up model to provide accurate wake simulation. Without the software, designers would have to guess the proper empirical constants in order to approximate wake characteristics.

According to CDI, the new fast vortex and fast panel technologies implemented within CHARM provide a great advancement in computational performance when compared to existing panel codes currently available.

Air Data Report Improves Flight Safety

Originating Technology/NASA Contribution

Aviation is one of the safest means of transportation, but aviation safety professionals always work to make it safer. When flights operate outside of the norm, analysts perk up, as these flights are perhaps also operating outside the realm of safety. These out-of-the-ordinary flights, or atypicalities, are, therefore, the ones that need to be studied, and this is where NASA steps in.

Traditionally, safety analysts compare data to preset parameters to determine the existence of atypical events, but a newly developed NASA program could point analysts to issues which might otherwise have been unforeseen if the analysts had only been looking for these predetermined events. The fundamental difference between NASA's methodology and traditional exceedance detection originates from the concept of detecting atypicalities without any predefined parameters.

This is the basic concept behind NASA's Morning Report software created at Ames Research Center. It is not the only software of its kind; rather, it aims to address some of the shortcomings of traditional safety systems. The software aggregates large volumes of flight data and then uses an advanced cluster-based, data-mining technique to find the unexpected or the abnormal, without needing the user to pre-define any events. Simply put, it spots deviations and highlights them for analysis.

The software was designed at Ames under the sponsorship of the Aviation Safety Program in the NASA Aeronautics Research Mission Directorate, which seeks to make aviation safer by developing advanced tools that find latent safety issues from large sources of flight digital and operational data sources.

Since its inception in 1999, the NASA team has collaborated with air carriers and vendors of flight operational quality assurance (FOQA) software, a widely used tool that seeks to provide airline managers with information that will enable them to better understand risks to flight operation and how to then manage these risks.

The focus on FOQA software led to development of the Morning Report tool. The tool, created with the assistance of the Pacific Northwest National Laboratory, provides flight analysts with a daily morning report of atypical flights, displayed with the ability to plot those parameters against what is typical for that phase of flight and particular airport.

The Morning Report tool uses multivariate statistical algorithms to analyze large amounts of data from airline flights overnight and then generates an intuitively struc-

tured report each morning. It combines these powerful algorithms for analysis with user-intuitive software, allowing users to isolate and understand details underlying any portion of any given flight. It is the only technology of its kind that provides both the global overview as well as the ability to view the smallest details of any flight.

Partnership

In 2004, Sagem Avionics Inc. entered a licensing agreement with NASA for the commercialization of the

Sagem Avionics incorporated NASA's Morning Report tool into its AGS comprehensive flight operations monitoring system, which processes all available data from aircraft recorders and provides customized reports.

Morning Report software. The company, based in Grand Prairie, Texas, also licensed the NASA Aviation Data Integration System (ADIS) tool, which allows for the integration of data from disparate sources into the flight data analysis process.

Sagem Avionics has been providing aeronautical equipment for flight testing, as well as acquisition, management, recording, and analysis of flight data for over 5 decades. In fact, Sagem Avionics' Analysis Ground Station (AGS) was the first commercially produced FOQA analysis tool. Similarly, it is the first and only analysis software that has incorporated the NASA Morning Report technology.

Product Outcome

Sagem Avionics' AGS product, incorporating the Morning Report tool, processes and analyzes available data from aircraft recorders and then produces easy-to-read, configurable, customized reports. The automated system is powerful enough to process very large volumes of data quickly and accurately, to help users detect irregular or divergent practices, technical flaws, and problems that might develop when aircraft operate outside of normal procedures.

In addition to the Morning Report technology, the AGS system also provides automatic, statistical, and manual analysis of flight data. The automatic analysis processes all available data from aircraft recorders and provides customized reports of daily events with classification levels. Since it is automated, the systematic rereading of recorded data minimizes repetitive daily tasks, and it updates its flight and event database regularly, making itself more knowledgeable and effective each day.

All events detected in the course of the automatic analysis are stored in the AGS database for statistical analysis, allowing users to produce predefined reports or create new ones to detect patterns and trends. These reports can be automatically edited, published, and exported in various formats, including HyperText Markup Language (HTML) and e-mail.

The manual analysis feature includes all of the necessary components for accurate investigations of any specific flight, allowing analysts to zero in on those flights operating outside of the norm. It displays engineering values in several formats (tables, curves, graphic charts, and generic cockpit instrument representation), making it ideal for investigating a wide range of these isolated events.

The fast and user-friendly AGS system manages large volumes and a wide variety of input data, but also monitors the media quality while controlling the entire data flow. The entire analysis process requires less than 3 seconds per flight hour and has been designed for compatibility with the standard personal computer. The complete system is integrated into a unique program with a standardized and homogeneous user interface.

The AGS turnkey system is plug-and-play, with all components integrated directly into the system. Users have access to decoding frames for aircraft parameter conversion in engineering units; procedure sets for the customer fleet and dedicated to flight operation and engineering maintenance analysis; and predefined statistic reports for periodic analysis of fleet activity.

To better meet customer requirements, Sagem developed the original AGS in collaboration with airlines, so that the system takes into account their technical evolutions and needs. Thanks to its modular architecture, AGS can be used by all carriers, from the smallest to the largest. Each airline is able to easily perform specific treatments and to build its own flight data analysis system. Further, the AGS is designed to support any aircraft and flight data recorders. ❖

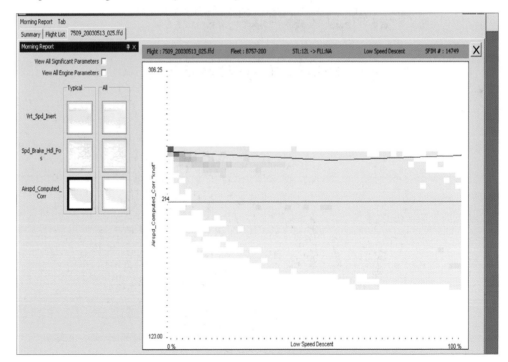

The Morning Report tool automatically identifies statistically extreme flights to airline FOQA analysts. The new software may help analysts identify the precursors of incidents or accidents.

Advanced Airfoils Boost Helicopter Performance

Originating Technology/NASA Contribution

Advanced rotorcraft airfoils developed by U.S. Army engineers working with NASA's Langley Research Center were part of the Army's risk reduction program for the LHX (Light Helicopter Experimental), the forerunner of the Comanche helicopter. The helicopter's airfoils were designed as part of the Army's basic research program and were tested in the 6- by 28-inch Transonic Tunnel and the Low-Turbulence Pressure Tunnel at Langley. While these airfoils did not get applied to the Boeing-Sikorsky Comanche rotor, they did advance the state of the art for rotorcraft airfoils.

The improved blade offered significantly greater lift capabilities, less drag, and less pitch (alternating lift and descent of the nose and tail of an aircraft during flight) than its predecessor and other conventional blades, when compared during high-speed flight-performance testing.

The Langley airfoil design, technically known as RC4, has managed to "lift off" and find much success in other applications.

Partnership

Carson Helicopters Inc. licensed the Langley RC4 series of airfoils in 1993—2 years after the Comanche project commenced—and began development of a replacement main rotor blade for their helicopters. The new Carson composite main rotor blades did not receive full Federal Aviation Administration certification until 2003. Regardless, it was well worth the wait for Frank Carson, president of the Perkasie, Pennsylvania-based company and a longtime helicopter pilot and designer. "The NASA Langley airfoil is one of the best airfoils in the world. Almost no one realizes how good it is. It's better than most anyone else has yet to come up with."

Product Outcome

Carson Helicopters provides a unique array of services that require hauling heavy loads. These services range from airlifting external cargo, suppressing wildfires, and carrying out emergency rescues, to performing high-rise rooftop installations, pouring concrete, and erecting steel structures and power lines in areas ground-based cranes cannot access (soft grounds and swamp areas, environmentally protected areas, and elevated/mountain terrain).

The company's entire fleet of Sikorsky S-61 helicopters has been rebuilt to include Langley's patented airfoil design on the main rotor. As a result of this retooling, the helicopters are now able to carry an additional 2,000 pounds (11,000 pounds total), fly 17 miles per hour faster, and travel 70 miles farther on the same fuel load. Additionally, the five blades on an S-61 main rotor are made from advanced composite materials that give them a

The NASA-developed rotorcraft airfoils permit greater lift and lower pitching-moments than other rotorcraft airfoils—ideal for heavy lifting operations.

The advanced airfoils, as applied to the main rotor of an S-61 helicopter, greatly increase the lifting capacity, operational envelope, and forward flight speed of the helicopter.

20,000-hour service life—double that of the conventional metal S-61 blades' life—which ultimately reduces the company's operating maintenance costs.

In 2003, Carson Helicopters signed a contract with Ducommun AeroStructures Inc., to have the Gardena, California-based firm manufacture the composite blades for Carson Helicopters to sell. The commercial blades are due to fly on a U.S. Presidential VH-3 helicopter in late 2007, as part of the VH-3D Lift Improvement Program sponsored by the U.S. Naval Air Systems Command. Overseas, a customized set of blades is being tested for potential use on a British Sea King Royal Navy helicopter.

In aerial firefighting, the performance-boosting airfoils have allowed for a major modification to be made to a Sikorsky S-61 chopper, in order to help the U.S. Department of Agriculture's Forest Service control the spread of wildfires. Because the airfoils permit heavier lifting than ever before, Carson Helicopters was able to install an oversized belly tank on its S-61 Fire King (or, as the company likes to call it, the "Swiss Army knife of helicopters"). With this modification, the Fire King is the only Type 1 helicopter (the heaviest type of helicopter, based on a scale of Type 1 to Type 3) that can carry up to 15 firefighters and hold 1,000 gallons of water in its belly tank at the same time. What's more, the helicopter has an 8,000-pound onboard cargo capacity. There are currently eight Fire Kings flying with the composite blades and the 1,000-gallon belly tank. ❖

Deicing System Protects General Aviation Aircraft

Originating Technology/NASA Contribution

Ice accumulation is a serious safety hazard for aircraft. The presence of ice on airplane surfaces prevents the even flow of air, which increases drag and reduces lift. Ice on wings is especially dangerous during takeoff, when a sheet of ice the thickness of a compact disc can reduce lift by 25 percent or more. Ice accumulated on the tail of an aircraft (a spot often out of the pilot's sight) can throw off a plane's balance and force the craft to pitch downward, a phenomenon known as a tail stall.

The Icing Branch at NASA's Glenn Research Center works using the Center's Icing Research Tunnel and Icing Research Aircraft, a DeHavilland Twin Otter twin-engine turboprop aircraft, to research methods for evaluating and simulating the growth of ice on aircraft, the effects that ice may have on aircraft in flight, and the development and effectiveness of various ice protection and detection systems.

Typically, ice is removed from general aviation craft with either "weeping wing" liquid deicing systems or inflatable rubber bladders, called pneumatic boots, installed along the wings. Both of these methods have drawbacks, including the finite, limited effectiveness of the liquid deicers and the added weight and power usage of the boots. Collaborative research at Glenn focused on using expanded graphite foil heating element technology to effectively replace these standard methods with a method that was usually limited to use on jets with heated wings and leading edge surfaces. The super-thin graphite, which covers a large surface area without significant weight penalties and heats quickly to melt ice, proved a viable solution, and this new safety equipment has now been made available to the aerospace community.

Partnership

Kelly Aerospace Thermal Systems LLC, of Willoughby, Ohio, is a division of Montgomery, Alabama-based Kelly Aerospace Inc., a leading subsystem supplier to general aviation equipment manufacturers and aftermarket customers. The Ohio-based design and development branch worked with researchers at Glenn on the deicing technology with assistance from the **Small Business Innovation Research (SBIR)** program.

Kelly Aerospace acquired Northcoast Technologies Ltd., a Cleveland-based firm that had similarly done graphite foil heating element work with NASA under an SBIR contract. Through its research, Northcoast had developed the Thermawing system, a lightweight, easy-to-install, reliable wing and tail deicing system. Kelly Aerospace engineers combined their experiences with those of the Northcoast engineers, and now continue to advance this work.

Product Outcome

The NASA-funded research has resulted in a handful of new products and applications, including the certification and integration of a thermoelectric deicing system, DC-powered air conditioning for single-engine aircraft, and high-output alternators to run them both.

Marketed as Thermawing, the aircraft deicing system employs a flexible, electrically conductive graphite foil

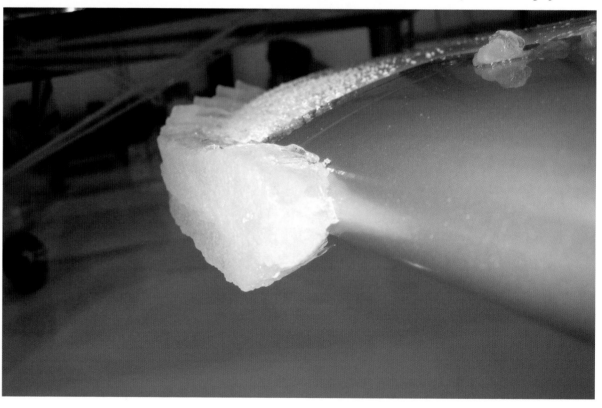

With NASA assistance, Kelly Aerospace has developed lightweight heating elements capable of keeping ice from forming on airplane wings.

that heats quickly for instantaneous rises in temperature when needed. It has an ultra-thin laminate construction that allows for low weight penalties. With this system, users are able to retrofit an aircraft with between 100- and 150-amp alternators producing 50 to 80 volts with negligible weight addition. This reliable anti-icing and deicing system allows pilots to safely fly through ice encounters and provides pilots of single-engine aircraft the heated wing technology usually reserved for larger, jet-powered craft.

It is simple to apply and requires far less wattage than standard electrical metal heating systems. The thin laminate system is applied like a tape, and it will bond to any surface of an aircraft where icing might become a problem. The laminate contains the flexible, expanded graphite foil that serves as an electrical and heat conducting layer, that works as effectively as multiple heat conducting layers and layers of electrical insulation. Energy can be controlled across the system, so that certain zones can be heated according to need, an energy-saving measure.

The Thermawing system is currently certified for use as airfoil protection on Columbia 350 and 400 single-engine aircraft, as well as the Beechcraft Baron B55. Kelly Aerospace is continuing to develop systems for other aircraft.

The company has also developed Thermacool, an innovative electric air conditioning system also for use on single-engine, general aviation aircraft. The typical method for cooling these aircraft uses a standard automobile air conditioner compressor, typically running off of a combination of belts hooked to the engine and electric motors, which drew too much energy for their use to be practical while on the ground or idling. Air conditioning in the cabin was available, then, only when the aircraft was airborne. Kelly Aerospace addressed this with a new compressor, whose rotary pump design runs off an energy-efficient, brushless DC motor. This now allows pilots to begin cooling the plane before the engine even starts.

Ice forming on an aircraft can pose serious risks. The Icing Branch at Glenn Research Center seeks to minimize these hazards by creating technologies that detect and prevent icing.

Weighing less than 14 pounds, the total system draws only 50 amps. The small compressor can be attached just about anywhere within the aircraft, and it is virtually maintenance free. Kelly Aerospace has been granted Supplemental Type Certificates (STCs) from the Federal Aviation Administration for use of the revolutionary air compressor on Cessna 182 models P, Q, and R, and Cessna 172 models R and S single-engine aircraft. It is currently developing a customized kit for the Piper PA-32 Cherokee Six, and more STCs are in the works.

To assist in running both the Thermawing deicing system and the Thermacool air conditioning system, Kelly Aerospace has designed an alternator capable of creating ample electricity, as well as the other complex electronics on the craft, whether the plane is airborne or idling on the ground.

Recently, Kelly Aerospace Thermal Systems entered into an agreement with Redmond, Oregon-based RDD Enterprises LLC, a developer of safety and performance systems for the experimental aircraft market. The partnership will allow Kelly Aerospace's thermal deicing systems to be widely available in this market. ❖

Thermawing™ and Thermacool™ are trademarks of Kelly Aerospace Inc.

Public Safety

NASA makes our world safer. The technologies featured in this section:

- Detect potential threats
- Sharpen views in critical situations
- Clean air and water for indoor environments

Chemical-Sensing Cables Detect Potential Threats

Originating Technology/NASA Contribution

As fleets of aircraft age, corrosion of metal parts becomes a very real economic and safety concern. Corrosive agents like moisture, salt, and industrial fluids—and even internal problems, like leaks and condensation—wear away and, especially over time and repeated exposure, begin to corrode aircraft.

Costs of repairs, which often involve replacing entire panels, downtime for fleet craft, and the potential loss of use of a vehicle all contribute to the economic concerns. In the United States alone, this is a multibillion dollar problem. In terms of safety, corrosion can be so widespread as to cause problems with the structural integrity of the craft, but it can also be just as dangerous when localized, like eating away at electrical wires or rusting away at crucial landing gear components.

NASA is invested in the development of new materials and material coatings to retard and prevent corrosion, but it is also researching methods for monitoring corrosion on existing aircraft.

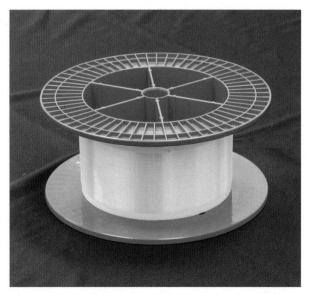

The DICAST system combines patented fiber optic cables that are chemically sensitive over their entire length, with an advanced network architecture for large-scale protection of critical assets.

Partnership

Moisture can lead to corrosion in an aircraft, and altered pH can indicate that corrosion is occurring. Intelligent Optical Systems Inc. (IOS), of Torrance, California, completed Phase I and II **Small Business Innovation Research (SBIR)** contracts with NASA's Langley Research Center to develop moisture- and pH-sensitive sensors to detect corrosion or pre-corrosive conditions. As opposed to typical inspection equipment, like ultrasound or X-ray imaging, this technique can warn of potentially dangerous conditions before significant structural damage occurs. This new type of sensor uses a specially manufactured optical fiber whose entire length is chemically sensitive, changing color in response to contact with its target. These sensor fibers, embedded directly in aluminum "lap joints," have detected the location—to within 2 centimeters—and alkalinity of potentially corrosive moisture incursions.

Product Outcome

After completing the work with NASA, the company received funding through a Defense Advanced Research Projects Agency (DARPA) Phase III SBIR to develop the sensors further for detecting chemical warfare agents. The sensors proved just as successful at locating and identifying specific chemicals as they had been at detecting moisture and pH levels. The next stage of development involved working with the U.S. Department of Defense (DoD) to fine-tune the sensors for detecting potential threats, such as toxic industrial compounds and nerve agents. Through the DoD contract, the company has set up beta testing in major metropolitan areas.

In addition to the work with government agencies, Intelligent Optical Systems has sold the chemically

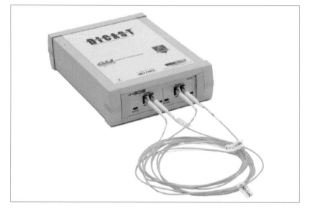

The sensor cables can be deployed in inexpensive alarm systems, which alert the user to a change in the chemical environment anywhere along the cable, or in sophisticated distance-resolved systems to provide detailed profiles of chemical concentration versus length.

sensitive fiber optic cables to major automotive and aerospace companies, who are finding a variety of uses for the devices, from the original corrosive-condition monitoring to aiding experimentation with nontraditional power sources.

Commercially marketed under the brand name DICAST (Distributed Intrinsic Chemical Agent Sensing and Transmission), these unique continuous-cable fiber optic chemical sensors use a glass core coated with a permeable indicator-doped cladding to achieve chemical sensitivity over their entire length. These fiber optic cables can be very long, providing an economical means of detecting chemical release in large facilities. They can be used in inexpensive alarm systems, which alert the user to a change in the chemical environment anywhere along the cable, or in distance-resolved optical time domain reflectometry systems to provide detailed profiles of chemical concentration versus length.

With a 10-second response time, the DICAST cables are able to detect agents used in chemical warfare, such as hydrogen cyanide and the nerve agents soman and sarin,

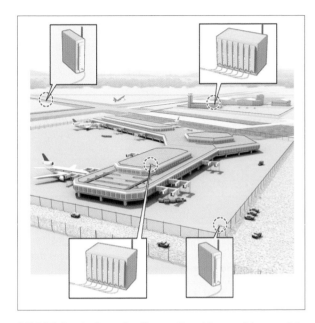

DICAST chemical-sensing fiber optic cables provide complete indoor and outdoor protection for large facilities.

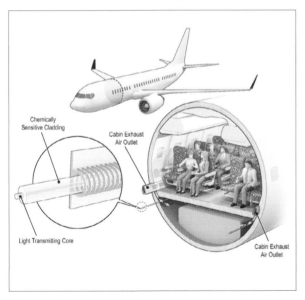

DICAST cable systems can be used on aircraft for passenger cabin protection.

as well as chlorine, hydrogen sulfide, and other industrial compounds. They are currently designed to operate in indoor environments and are intrinsically sensitive to select, individual chemicals. When any area along the length of a cable is exposed to the target chemical, the cable color changes, and a proprietary optoelectric system detects the change in light and alerts the user. The cable is sheathed in an air-permeable casing, which provides rugged protection against shear stress.

Uses for the DICAST cables abound in security and industrial applications. They can be used to protect buildings and structures, like subway stations, automotive tunnels, high-rise office buildings, arenas, and convention centers, when employed in ventilation ducts, stairwells, or even in large open spaces. The reduction in complexity and increased security provided by these continuous-cable chemical detection sensors make the DICAST system an ideal choice for these applications. A single cable running through a main return air duct, for example, is a more cost effective and efficient method for chemical detection than the installation of point sensors in each room's inlet to that duct.

In addition to their alarm and protection capabilities, the cables are also ideal for collecting profiles of chemical agents after an incident and for monitoring decontamination. By setting up a grid of cables over a contaminated area, response and clean-up crews can generate an accurate picture of the types, location, and concentration of hazardous chemicals.

Similarly, the DICAST cables can also be used to effectively secure a perimeter or boundary, with a single sensor cable replacing dozens of point sensors with an effectiveness that could only be rivaled by placing the point sensors side-by-side along the length of the entire perimeter. Possible locations that could benefit from this application include airbases, ports, high-profile

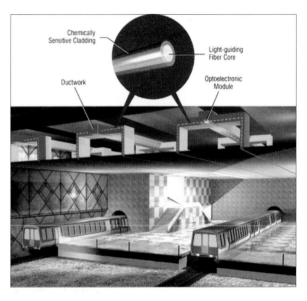

The cable systems can be used for building and structure protection, such as in ventilation ducts, stairwells, and large open spaces in subways, high-rise buildings, arenas, convention centers, and automotive tunnels.

government installations, and facilities that either store or destroy chemical warfare agents. According to the company, the cables could also be used to create a ring of protection around large outdoor gatherings.

Industrial applications, in addition to the original uses of moisture and corrosion detection in aircraft, include detection of non-lethal chemicals, such as leaks in long-haul water or petrochemical pipelines, pipes, and fluid transfer lines, and in refineries and other industrial facilities. The moisture and corrosion detection, though originally developed for aircraft applications, can also be applied to ships and other structures, and can even be used to detect salt and moisture permeation in concrete. ❖

DICAST® is a registered trademark of Intelligent Optical Systems Inc.

Infrared Imaging Sharpens View in Critical Situations

Originating Technology/NASA Contribution

The Microgravity Combustion Science group at NASA's Glenn Research Center studies how fire and combustible liquids and gasses behave in low-gravity conditions. This group, currently working as part of the Life Support and Habitation Branch under the Exploration Systems Mission Directorate, conducts this research with a careful eye toward fire prevention, detection, and suppression, in order to establish the highest possible safety margins for space-bound materials.

Over the years, the group has established that many materials burn very differently in microgravity than they do on Earth. For example, attempting to stomp out a flame in microgravity could possibly accelerate combustion, at least temporarily (because an airflow is being created that did not exist before). Other interesting findings indicate that microgravity fires can spread faster upstream than downstream, opposite of the behavior of fire spreading on Earth, and that fire is actually weaker in microgravity. As a matter of fact, the weakest flames ever generated were done so in space. During the "Structure of Flame Balls at Low Lewis-number" (SOFBALL) experimental trials conducted during missions STS-83 (April 1997), STS-94 (July 1997), and STS-107 (January 2003), flames were generated in space with power as low as 1 watt—about 50 times weaker than a candle flame.

This is not to say that fire is safer in space, though. Fire outbreak on a spacecraft is just as dangerous as any fire situation on Earth, or arguably even more dangerous, given the inability of astronauts to evacuate. For this reason, the ability to detect subtle variations in temperature in a complex and varied thermal background could prove invaluable in a spacecraft.

Partnership

Innovative Engineering and Consulting (IEC) Infrared Systems is a leading developer of thermal imaging systems and night vision equipment. The Cleveland-based company was founded in 1999 by two microgravity combustion science researchers from the National Center for Space Exploration Research, an academic research organization located onsite at Glenn. In spinning off their new business venture, the two researchers utilized the engineering know-how they developed in measuring high-temperature flames for NASA space flight experiments.

Several years after opening for business, IEC Infrared Systems received a Glenn Alliance for Technology Exchange (GATE) award worth $100,000, half of which was in the form of additional NASA assistance for new product development. The GATE award was established by Glenn, the Ohio Aerospace Institute, and the Battelle Memorial Institute to assist small Ohio-based companies interested in collaborating with NASA to advance their products and processes.

IEC Infrared Systems used the funds earmarked for NASA assistance to work with electrical and optical engineers from Glenn's Diagnostics and Data Systems Branch on the development of a commercial infrared imaging system that could differentiate the intensity of heat sources better than other commercial systems. Firefighters, for example, could use the proposed technology to make clearer distinctions between the intense heat of a fire and the lower-level thermal signatures of human bodies in fire-based search and rescue situations where darkness, smoke, or fog can obscure their vision.

Product Outcome

The firsthand NASA knowledge and the follow-up funding and technical support from the GATE award were the catalysts for IEC Infrared Systems to evolve from a start-up venture to a multimillion dollar business with a staff of more than 30 scientists, engineers, and technicians spanning a wide range of engineering fields. Today, the company offers two major thermal imaging solutions that stem directly from its vast NASA experience: NightStalkIR and IntrudIR Alert ("IR" for infrared). These two imaging systems have found widespread use

IEC Infrared Systems' Thermal/Visual Imaging Systems product line features a suite of thermal imaging devices that are used in a wide range of applications.

in emergency first-responder, facility security, and military applications.

With advanced imaging capabilities, proprietary signal processing and electronics, and a tough, rugged exterior, NightStalkIR offers optimal daytime/nighttime surveillance in all weather conditions. Features include a low-light camera with lens options ranging from 50 to 180 millimeters (without the "halo" effect commonly seen in some imagers), full 360-degree rotation with pan and tilt, hand controller and PC software control, fixed or mobile (vehicle) mounting, and onscreen positional display of imaging direction and other tactical data. Optional features include fiber optic and wireless capabilities, an image-intensified camera that further enhances nighttime imaging in the visual spectrum, and Global Positioning System/compass/laser range systems that provide the precise location of observed targets for increased tactical awareness.

IntrudIR Alert is an intrusion-detection system designed to operate with multiple NightStalkIR thermal imagers. This software-based system allows a single operator to command and control these imagers over a broad area, for maximum tactical and situational awareness and early warning of intrusions. Features include target tracking, based on either thermal signatures or motion detection (or a combination of both), continuous automatic tracking of these targets, and digital capturing of still images or short video clips, either on command or in response to alarms. In addition to being compatible with NightStalkIR, IntrudIR Alert can be integrated into larger security networks.

According to IEC Infrared Systems, NightStalkIR and IntrudIR Alert are being used in the United States and abroad to help locate personnel stranded in emergency situations, defend soldiers on the battlefield abroad, and protect high-value facilities and operations. The company is also applying its advanced thermal imaging techniques to medical and pharmaceutical product development with a Cleveland-based pharmaceutical company. This cooperative effort was enabled by a NASA Space Act Agreement, as Glenn continues to encourage IEC Infrared Systems' founding partners to explore new product ideas based on the techniques developed during their tenure at NASA. For the founders, their work with NASA and their related commercial endeavors have given a whole new meaning to "playing with fire." ❖

NightStalkIR™ and IntrudIR Alert™ are trademarks of Innovative Engineering and Consulting Infrared Systems.

Modern infrared imaging devices allow users to see through smoke and fog, as well as in total darkness. They have widespread use in military, facility security, and emergency first-responder applications.

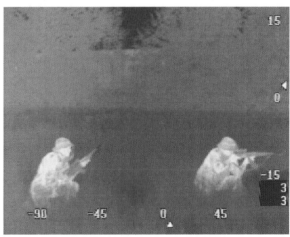

This software-based system allows a single operator to command and control these devices over a broad area, for maximum tactical and situational awareness and early warning of intrusions.

Plants Clean Air and Water for Indoor Environments

Originating Technology/NASA Contribution

Although one of NASA's goals is to send people to the far reaches of our universe, it is still well known that people need Earth. We understand that humankind's existence relies on its complex relationship with this planet's environment—in particular, the regenerative qualities of Earth's ecosystems.

In the late 1960s, B.C. "Bill" Wolverton was an environmental scientist working with the U.S. military to clean up the environmental messes left by biological warfare centers. At a test center in Florida, he was heading a facility that discovered that swamp plants were actually eliminating Agent Orange, which had entered the local waters through government testing near Eglin Air Force Base. After this success, he wanted to continue this line of research and moved to what was at the time called the Mississippi Test Facility, but is now known as NASA's Stennis Space Center.

He was funded by the Space Agency to research the environment's natural abilities to clean itself as part of what is now Stennis' Environmental Assurance Program. The goals were to clean the Center of chemicals left behind through wastes and to supply information to NASA engineers about closed-environment "eco" support that may prove helpful in designing sustainable living environments for long-term habitation of space. A tertiary goal was to provide usable technologies to NASA's Technology Utilization Program, essentially making the research available to the American public.

The first step for Wolverton's research was to continue the remediation work he had started with the military. He was tasked with using plants to clean waste water at the NASA Center. To this day, Wolverton's design, which replaces a traditional septic system with water hyacinths, is still in use. His research then turned to using plants to improve air quality.

In 1973, NASA scientists identified 107 volatile organic compounds (VOCs) in the air inside the Skylab

The BioHome at NASA's Stennis Space Center was 45 feet long, 16 feet wide, and used common indoor house plants as living air purifiers.

space station. Synthetic materials, like those used to construct Skylab, give off low levels of chemicals. This effect, known as off-gassing, spreads the VOCs, such as formaldehyde, benzene, and trichloroethylene, all known irritants and potential carcinogens. When these chemicals are trapped without circulation, as was the case with the Skylab, the inhabitants may become ill, as the air they breathe is not given the natural scrubbing by Earth's complex ecosystem.

Around the same time that Wolverton was conducting his research into VOCs, the United States found itself in an energy crisis. In response, builders began making houses and offices more energy efficient. One of the best ways to do this was to make the buildings as airtight as possible. While keeping temperature-controlled air in place, this approach reduced circulation. Combined with the modern use of synthetic materials, this contributed to what became known as Sick Building Syndrome, where toxins found in synthetic materials become concentrated inside sealed buildings, making people feel sick.

The solution Wolverton sought was not to make indoor environments less energy efficient or to move away from the convenience of synthetic materials; rather, the plan was to find a solution that restores personal environments. The answer, according to a NASA report later published by Wolverton in 1989, is that "If man is to move into closed environments, on Earth or in space, he must take along nature's life support system." Plants.

One of the NASA experiments testing this solution was the BioHome, an early experiment in what the Agency called "closed ecological life support systems." The BioHome, a tightly sealed building constructed entirely of synthetic materials, was designed as suitable for one person to live in, with a great deal of the interior occupied by houseplants. Before the houseplants were added, though, anyone entering the newly constructed facility would experience burning eyes and respiratory difficulties, two of the most common symptoms of Sick Building Syndrome. Once the plants were introduced to the environment, analysis of the air quality indicated that most of the VOCs had been removed, and the symptoms disappeared.

Partnership

After serving over 30 years as a government scientist, Wolverton retired from civil service but continued his work in air and water quality by founding Wolverton Environmental Services Inc. The company, based just down the road from Stennis in Picayune, Mississippi, is an environmental consulting firm that gives customers access to Wolverton's decades of cutting-edge bioremediation research.

Product Outcome

Wolverton published his findings about using plants to improve indoor air quality in dozens of technical papers while with the Space Agency and as a simple consumer-friendly book, "How to Grow Fresh Air: 50 Houseplants That Purify Your Home or Office." In it, he explains, in easy-to-understand terms, how plants emit water vapor that creates a pumping action to pull contaminated air down around a plant's roots, where it is then converted into food for the plant. He then goes on to explain which plants and varieties remove the most toxins, as well as to rate each plant for the level of maintenance it requires. The book has now been translated into 12 languages and has been on the shelves of bookstores for nearly 10 years. Wolverton has also published a companion book, "Growing Clean Water: Nature's Solution to Water Pollution," which explains how plants can clean waste water.

Another one of Wolverton's discoveries is that the more air that is allowed to circulate through the roots of the plants, the more effective they are at cleaning polluted air. To take advantage of this science, Wolverton has teamed with the Japanese company, Actree Corporation, to develop what the Japanese firm is marketing as the EcoPlanter. Using high-efficiency carbon filters and a root-level circulation system, the pot allows the plant to remove approximately 200 times more VOCs than a single traditionally-potted plant can remove.

The company has recently begun to assess the ability of the EcoPlanter to remove formaldehyde from the many travel trailers furnished by the Federal Emergency Management Agency to victims of Hurricane Katrina. The interiors of the trailers make heavy use of particleboard, which off-gasses formaldehyde. Many of the trailers have been found to exceed the recommended levels of formaldehyde for human safety. Initial tests of the EcoPlanter have been very encouraging, but other testing is still needed.

Research has also suggested that plants play a psychological role in welfare, and that people actually recover from illness faster in the presence of plants. Wolverton's company is working with another Japanese company, Takenaka Garden Afforestation Inc., of Tokyo, to design ecology gardens. These are carefully designed gardens that help remove the toxins from the air in hospitals, as well as provide the healing presence of the foliage.

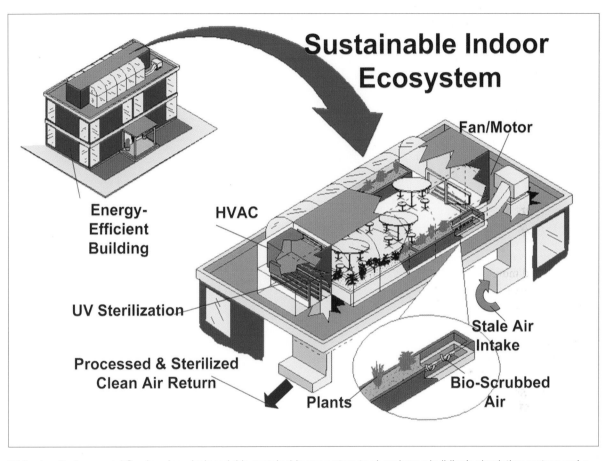

Wolverton Environmental Services Inc. designed this sustainable ecosystem to show how a building's circulation system and a rooftop garden could work in tandem to clean indoor air.

On the home front, in a partnership with Syracuse University, Wolverton Environmental is engineering systems consisting of modular wicking filters tied into duct work and water supplies, essentially tying plant-based filters into heating, ventilation, and air conditioning (HVAC) systems. This whole-building approach has recently been licensed by Wolverton to Phytofilter Technologies Inc., of Saratoga Springs, New York, which is currently constructing a prototype of a system that is intended to clean the water and air circulation systems of entire buildings using the natural abilities of plants. The design includes units that are built into existing HVAC units. The plants can be placed throughout buildings, in atriums, or in roof gardens and then hooked into the building's HVAC units through forced-air filters.

Wolverton Environmental is also in talks with designers of the new Stennis Visitor's Center, who are interested in using its designs for indoor air-quality filters. ❖

Consumer, Home, and Recreation

NASA research improves our quality of life. The technologies featured in this section:

- Restore artwork
- Enhance education and recreation
- Reduce fat while improving flavor
- Transform paint into insulation
- Protect machines and the environment

Corrosive Gas Restores Artwork, Promises Myriad Applications

Originating Technology/NASA Contribution

Short wavelength solar radiation in the space environment just outside of the Earth's atmosphere produces atomic oxygen. This gas reacts with spacecraft polymers, causing gradual oxidative thinning of the protective layers of orbiting objects, like satellites and the International Space Station, which maintain low-Earth orbit directly in the area where the corrosive gas is most present.

To combat this destructive gas, NASA engineers developed long-duration coatings that are resistant to the effects of its problematic presence. To validate the effectiveness of the coatings, NASA had two options: Either send the materials into orbit for testing, which would involve the cost of launches and severely limit access to the experiments, or recreate the atmospheric conditions here on Earth. NASA chose the latter, and the Electro-Physics Branch at Glenn Research Center constructed ground facilities to test the durability of different materials by introducing them to a recreated form of the corrosive space gas.

The experiments were successful, and the coatings are currently used on the International Space Station. In the experimentation, though, the scientists discovered several additional interesting applications for their test facilities and beneficial uses for atomic oxygen here on Earth.

Partnership

Led by Glenn's Bruce Banks and Sharon Rutledge, the Electro-Physics researchers became familiar with atomic oxygen's unique characteristic of oxidizing hydrogen, carbon, and hydrocarbon polymers at surface levels. While destructive to spacecraft polymers constructed with those materials, atomic oxygen's selectivity could, they realized, also be applied in instances where someone wanted just those elements removed. Over the past few years since they made this realization, Banks and his team

The Atomic Oxygen Exposure Facility in operation at Glenn Research Center has been used for the removal of smoke damage and aged varnish from the surface of paintings and for cleaning organic contaminants from surfaces of materials.

have partnered with several churches and museums to restore fire-damaged or vandalized artworks, and with an international forensics organization to develop new methods for detecting forged documents, as well as having developed a method for using atomic oxygen to remove bacterial contaminants from surgical implants.

Product Outcome

Atomic oxygen is able to remove organic compounds high in carbon (mostly soot) from fire-damaged artworks without causing a shift in the paint color. It was first tested on oil paintings. In 1989, an arson fire at St. Alban Episcopal Church, in Cleveland, nearly destroyed a painting of Mary Magdalene. Although the paint was blistered and charred, after 230 hours of atomic oxygen treatment and a reapplication of varnish, it was once again recognizable as a work of art. In 2002, a fire at St. Stanislaus Church, again in Cleveland, left two paintings with soot damage that the atomic oxygen process was able to remove.

Buoyed by the successes with oil paints, the team also applied the restoration technique to acrylics, watercolors, and ink. As long as the paints were primarily synthetic, the results were promising. They discovered though, that some organic acrylics and ink, in particular, required less exposure so that the atomic oxygen would not begin to wear away at the medium itself. This potential liability has been used advantageously, however, in instances of graffiti removal. Experiments showed that, by using a pencil-thin beam of atomic oxygen, the team was able to remove most inks except black permanent marker.

At Pittsburgh's Carnegie Museum of Art, where an Andy Warhol painting, "Bathtub," was kissed by a lipstick-wearing vandal, the technique successfully removed the offending pink mark with a portable atomic oxygen gun. The process lightened a spot of paint, but a conservator was easily able to match the spot, thus restoring the painting.

The successes with the art restoration process were well-publicized, and Lynda Taylor-Hartwick of the Independent Association of Questioned Document Examiners Inc. (IAQDE), a multinational, nonprofit professional organization dedicated to the art of forensic analysis of documents, read about the effects of atomic oxygen on ink and became curious about possible applications for this process in the field of forgery detection. She found that it can assist document analyzers in determining if, for example, checks or wills have been altered.

Atomic oxygen oxidation of ink may cause altered pen marks to look differently than the original marks. It can help examiners discriminate between two different inks, because different inks may oxidize at different rates, showing document examiners any signs of tampering. Usefulness, however, is not limited to instances where the inks are of different manufacture. Atomic oxygen, which oxidizes and removes organic materials by converting them into gasses, works gradually. Thus, thick layers of carbon or organic materials take longer to remove than thin layers. The ends of pen strokes tend to have much thicker ink deposits than the rest of the line, enabling the use of atomic oxygen exposure to determine which lines were drawn first, which strokes were made as one fluid movement, and which overlapped strokes have been added at a later date, a clear indication that a document has been altered.

The most telling sign, though, is the layering of ink that occurs when someone writes over a letter or number to alter it. Take, for example, the classic case of modifying a report card to turn an F into a B before showing the parents. To complete this feat, the belatedly concerned student would connect the lines at the top of the F with a curved stroke, making it more similar to the letter P, and then finish the job by looping in the base, thus raising the grade to a B. In order to make the job look good, though, the strokes must connect to the original letter, even overlap a little to make it look uniform. It is the overlapping, a miniscule amount of layering, that atomic oxygen can erode in order to expose the alteration.

While most parents may not go the extent of acquiring a portable atomic oxygen gun to check a report card, the application becomes more relevant for applications like determining check fraud or altered wills. Just as an F can become a B, a 1 can become a 9 or a 3 can become an 8, which could have potentially significant financial implications in instances of fraud.

Parishioners at St. Alban Church, in Cleveland, thought this painting of Mary Magdalene was ruined after an arson fire destroyed much of the property. The same corrosive attributes of atomic oxygen that eat away at spacecraft were able to remove the layers of soot and smoke that covered the painting.

It is not just paint and ink that the Glenn team is experimenting on, though. The gas has biomedical applications as well. Atomic oxygen technology can be used to decontaminate orthopedic surgical hip and knee implants prior to surgery. As a result of handling, fabrication, and exposure to air, the surfaces of these implants are often contaminated with endotoxins (naturally occurring compounds found within bacteria) and other biologically active contaminants. Such contaminants contribute to inflammation, which can lead to joint loosening, pain, and even the necessity to remove the implant. Previously, there was no known chemical process which fully removed these inflammatory endotoxins without damaging the implants. Atomic oxygen, however, can oxidize endotoxins and any other organic contaminants to convert them into harmless gasses, leaving a contaminant-free surface.

The inventors have patented this application for atomic oxygen and believe it could lead to significant reduction in health care costs for the more than 2.8 million people who receive orthopedic implants annually. They also believe that it promises increased functional life of implants, as well as a reduction of inflammation and the associated joint pain that patients experience.

Additional collaborative research between the Cleveland Clinic Foundation and the Glenn team into the terrestrial uses of atomic oxygen shows that this gas's roughening of surfaces even improves cell adhesion, which is important for the development of new drugs.

While this application is still in its testing stages, the others are available for use. The patent for atomic oxygen art restoration is now in the public domain. Use of the technology for document alteration detection was never patented, and it, too, is available in the public domain. A patent was licensed for the removal of biologically active components from surgical implants, and Glenn is currently in talks with a company that sells plasma treating equipment. ❖

Detailed Globes Enhance Education and Recreation

Originating Technology/NASA Contribution

Earth from space—swirling wisps of white against a backdrop of deep azure, punctuated with brown and green swatches of land, all etched on one orb surrounded by black space, floating, seemingly isolated, but teeming with humanity and other forms of life. It is an iconic image, first captured November 10, 1967, by the Applications Technology Satellite (ATS)-3, an unmanned craft conducting payload experiments and examining the space environment. Since then, astronauts and spacecraft have sent back hundreds of pictures of Earth, and each one has had the same breathtaking effect.

Seeing our home planet from space is one of those self-reflective experiences, like seeing yourself in a picture, or hearing your voice on tape. It tells you something about yourself from outside of yourself. It is an experience that changes your understanding of the world and your place in it.

This phenomenon is best illustrated by the words of space travelers who, upon reaching orbit, have gazed back at Earth and felt the profound impact of viewing the planet in its entirety.

Frank Borman, Apollo 8 commander, said, "The view of the Earth from the Moon fascinated me—a small disk, 240,000 miles away. . . . Raging nationalistic interests, famines, wars, pestilence don't show from that distance."

Another veteran of the Apollo 8 mission, William Anders, had this to say: "We came all this way to explore the Moon, and the most important thing is that we discovered the Earth."

Neil Armstrong, the first person to step foot on the Moon, described the feeling of perspective he experienced when staring out at the Earth from the spacecraft window: "It suddenly struck me that that tiny pea, pretty and blue, was the Earth. I put up my thumb and shut one eye, and my thumb blotted out the planet Earth. I didn't feel like a giant. I felt very, very small."

Alan Shepard, commander of the Apollo 14 mission, the eighth manned mission to the Moon, said of the experience of seeing the home planet in its entirety, "If somebody had said before the flight, 'Are you going to get carried away looking at the Earth from the Moon?' I would have said, 'No, no way.' But yet when I first looked back at the Earth, standing on the Moon, I cried."

Apollo 15 astronaut, James Irwin, said of the experience, "As we got further and further away, it [the Earth] diminished in size. Finally it shrank to the size of a marble, the most beautiful you can imagine. That beautiful, warm, living object looked so fragile, so delicate, that if you touched it with a finger it would crumble and fall apart. Seeing this has to change a man."

Astronaut Alfred Worden, another of the original Apollo crew, and pilot of the Apollo 15 mission, said of his experience, "Now I know why I'm here. Not for a closer look at the Moon, but to look back at our home, the Earth."

Years later, space shuttle astronauts are still finding the view of Earth from space a humbling and profound experience. Don Lind, astronaut aboard the STS-51B mission

This 16-foot-diameter, pedestal-mounted, rotating globe was featured at the 2006 Wirefly X-Prize Cup in Las Cruces, New Mexico. This exposition of personal space flight is a "celebration of forward-looking technology, space exploration, and education," attracting thousands of people every October.

said, "Think about the picture of the Earth coming up over the horizon of the Moon, which I call the picture of the century. Every crew that went to the Moon took that same picture. It is probably as moving as anything in reorienting people's idea to the whole world concept."

The experience was perhaps best illustrated, however, by Sultan bin Salman bin Abdulaziz Al Saud, a Saudi Arabian payload specialist aboard STS-51G, a space shuttle mission comprised of American, French, and Saudi astronauts, when he said, "The first day or so, we all pointed to our countries. The third or fourth day, we were pointing to our continents. By the fifth day, we were aware of only one Earth."

Partnership

In 1985, inspired by the pictures he had seen of the Earth from space, Eric J. Morris founded a company devoted to designing and producing photorealistic replicas of our planet. The company, Orbis World Globes, is located in Eastsound, Washington, and creates inflatable globes in many sizes that depict Earth as it is seen from space, complete with atmospheric cloud cover.

Utilizing available technology of the mid-1980s, the original EarthBall design was derived from an artist's compilation of NASA photographs and National Oceanic and Atmospheric Administration (NOAA) weather satellite faxes. Over the following 2 decades, as 3-D imaging software, more powerful computers, and superwide inkjet printers became readily available, Orbis employed these evolving capabilities to develop larger and more accurate world globes. Applying proprietary cartographic software to transform flat-map imagery into multiple gores (2-D tapered sections) for creating spherical globes, Orbis now designs and produces the most visually authentic replicas of Earth ever created.

NASA took notice of Orbis globes and employed a 16-inch-diameter EarthBall for an educational film it made aboard the STS-45 shuttle mission. "The Atmosphere Below," shot during the 9-day mission

This 6-foot-diameter rotating Orbis globe is installed in the Children's Wing Atrium of Northland Church in Longwood, Florida.

aboard Atlantis in 1992, features astronauts using the EarthBall to explain how scientists use the unique vantage point of space to study the Earth's atmosphere.

Orbis later collaborated with NASA to create two giant world globes for display at the 2002 Olympic Winter Games in Salt Lake City. Creating the two 16-foot-diameter Earth globes with NASA involved updating the lower-resolution imagery the company had been using with more detailed, recent satellite imagery. This imagery was taken by NASA's Moderate Resolution Imaging Spectroradiometer satellite, combined with

observations of Antarctica made by NOAA's Advanced Very High Resolution Radiometer sensor.

Though the cloud cover has been slightly reduced in order for most of the landforms to be visible, Orbis globes are otherwise meteorologically accurate. In the Northern Hemisphere, it is fall, and it is spring in the Southern Hemisphere.

The satellite image now printed on all Orbis globes displays 1-kilometer resolution, which means that each pixel in the digital image represents 1 square kilometer of the planet's surface. It is 21,600 by 43,200 pixels in size. According to Orbis, while specific portions of the Earth have been imaged with higher detail, as of 2002, the satellite image it uses was the most detailed composite image of the entire world in existence.

NASA's new Blue Marble: Next Generation satellite imagery now provides even higher 500-meter resolution, which Orbis is planning to utilize for giant (30-foot plus) diameter world globes.

Orbis also developed the exclusive NightGlow Cities feature, enabling EarthBalls to display the world's cities as the globes revolve from daylight into night. Using light emissions data from U.S. Department of Defense satellites, city lights are identified with photoluminescent ink, fluorescing brightly under black ultraviolet light, adding a new dimension of authenticity to these world replicas.

Product Outcome

Orbis inflatable globes are available in sizes from 1 to 100 feet in diameter, with the most common being the standard 16-inch and 1-meter diameter EarthBalls. These smaller globes are ideal for educational purposes and have been used everywhere from preschools to universities. They come with a 20-page book of facts, games, and suggested activities, including a game of indoor classroom volleyball. They have been sold in thousands of gift shops, toy stores, museum shops, and specialty stores.

Over 100 Orbis world globes from 2 to 20 feet in diameter have been custom-built for a variety of display purposes. They have been used as temporary exhibits at numerous events, such as conferences, trade shows, festivals, concerts, and parades. For these types of purposes, the company maintains a fleet of rental globes ranging from 3 to 16 feet in diameter. Globes of various sizes have been exhibited at numerous events

This un-retouched photograph of a 4-foot-diameter Orbis globe features the world's cities fluorescing under black ultraviolet light. It was printed on special blackout digital fabric with the city lights added using ultraviolet-sensitive fluorescent paint.

Eric J. Morris, founder and chief cartographer of Orbis World Globes with a 4-foot-diameter cloud-free globe with the imprinted logo of the Instituto Nacional De Technica Aerospacial (National Institute of Aerospace Technology) in Spain.

worldwide. Last year, a 16-foot-diameter Orbis globe was exhibited at the United Nations' World Urban Forum, in Vancouver, Canada; the Space 2006 conference, in San Jose, California; and the X-Prize Cup Personal Spaceflight Exposition, in Las Cruces, New Mexico.

Giant Orbis globes have been put on permanent display at schools, churches, museums, libraries, and a variety of other locations. Orbis recently installed a 10-foot-diameter, internally illuminated, rotating globe in the education center at Washington's Fairchild Air Force Base; an 8-foot-diameter rotating world globe at the entrance to the new Evolving Planet exhibit in Chicago's Field Museum; and another 8-foot-diameter rotating world globe with NightGlow Cities in the new Ripley's Believe It or Not! museum in New York City's Times Square.

Orbis globes can be mounted on floor pedestals, suspended overhead, filled with helium and tethered

This 10-foot-diameter Orbis globe is being carried in a peace march in Seattle, Washington, in March 2005. Orbis globes have been a magnet for spectators and cameras in numerous parades, performances, and rallies around the world. (The gold building in the background is the Experience Music Project designed by renowned architect Frank Gehry.)

in the air, or even deployed as remote-controlled flying "EarthBlimps." Small electric motors enable the globes to slowly rotate on their axes, as does the real Earth. Internal illumination, custom graphics, and many other options are also available.

Twenty-two years ago, Eric J. Morris mused, "If only others could see Earth as the astronauts observe it, perhaps they would be similarly inspired." Considering the success of Orbis World Globes, people of all ages are being inspired everyday to appreciate our wondrous planet, Earth. ❖

Food Supplement Reduces Fat, Improves Flavor

Originating Technology/NASA Contribution

During the Mercury missions, astronauts ate terrible food: freeze-dried powders and semi-liquids in aluminum tubes. Decades later, though, astronauts now have meals prepared by celebrity chefs and access to everyday items like shrimp cocktail, stir-fried chicken, and fettuccine alfredo. While the culinary selection has improved, the developers of these gourmet delights are still faced with a number of challenges.

Space foods, which can be available in rehydratable, thermostabilized, irradiated, and natural forms, are tested for their nutritional value, appeal to the senses, storability, and packaging. The foods are also tested to ensure that they are low in weight and mass, require little energy to prepare for eating, have a minimum of 9 months shelf life for shuttle missions and 1 year for use on the International Space Station (ISS), and are stored at room temperature.

Additional challenges include the need to develop foods and equipment that take up very little space, are easy to operate and clean, and require minimal water use, while also creating minimal air pollution and odors, which can be hazardous to the health and well-being of astronauts. The foods must be crumb-free to eliminate excess floating particles. Space foods must also be free of pathogenic microbes and create minimum waste and mess.

Finally, space foods have to taste good, while still managing to be healthy. Toward this effort, NASA testing helped in the development of a revolutionary new fat substitute that cuts calories and extends shelf life.

Partnership

The NASA Glenn Garrett Morgan Commercialization Initiative (GMCI) is a program for small, minority-owned, and woman-owned businesses that can benefit from access to NASA resources. GMCI provides services that enable companies to grow and strengthen their business by leveraging NASA technology, expertise, and programs.

Diversified Services Corporation, a minority-owned business based out of Cleveland, Ohio, was able to take advantage of this NASA program for technology acquisition and development, and for introductions to potential customers and strategic partners, such as the NASA Food Technology Commercial Space Center, at Iowa State University (the center closed December 31, 2005), for taste tests and performance studies. Fresh ground beef (90-percent lean) was used to prepare hamburger patties formulated with or without 10-percent fat substitute. Hamburger patties without the added fat substitute served as the control in each experiment. Patties were weighed for evaluation of cooking yield, and then cooked to an internal temperature of 72 °C. The cooked product with or without fat substitute was rapidly cooled, and then subjected to freeze drying or irradiation in retort pouches to NASA specifications. Changes in volatile profile during storage, and sensory properties were determined. Addition of 10-percent fat substitute did not influence the sensory characteristics of the ready-to-eat hamburger beef patties or dramatically change its volatile profile after 30-day storage.

With the GMCI assistance, the company developed and commercialized a new nutritional fat replacement and flavor enhancement product it had licensed from the U.S. Department of Agriculture and is now marketing it through its subsidiary, H.F. Food Technologies Inc.

Product Outcome

The Nutrigras fat substitute is available for commercial applications and helps to satisfy the body's desire for the taste and mouth feel of fatty foods, even though the body does not actually need these foods—in fact, many people need fewer high-fat foods in their diets. With obesity on the verge of outweighing smoking as the number one cause of preventable death, the Centers for Disease Control and Prevention are showing rapid rises in the prevalence of

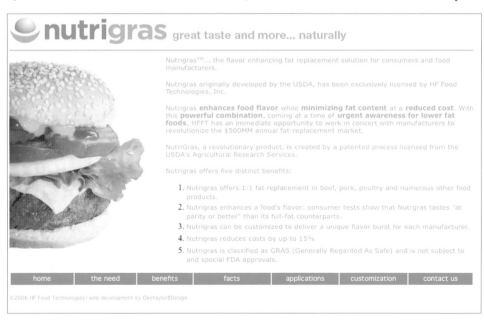

Diversified Services Corporation developed and commercialized a new nutritional fat replacement and flavor enhancement product with assistance from NASA.

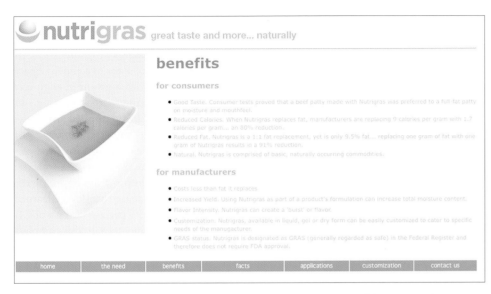

The Nutrigras food supplement creates food that is more moist, more tender, and more flavorful than its full-fat counterpart, and 1 pound of the supplement replaces 1 pound of animal fat.

children at severe weight levels; and while the American diet continues to be reliant on large quantities of high-fat foods, nutritionists are searching for solutions.

Nutrigras is one such solution. It is a stable emulsion of 9-percent vegetable oil and 62-percent water that has been formed by turbid excess steam jet cooking and made stable with microencapsulation in a starch solution that holds the emulsion steady. It is available in liquid, gel, or dry form and can be easily customized to the specific needs of the food manufacturer. When constituted, it looks and tastes just like real fat, but it is significantly healthier.

It is primarily intended for use as a partial replacement for animal fat in beef patties and other normally high-fat meat products, but can also be used in soups, sauces, bakery items, and desserts.

Nutrigras is intended to be used as a direct, pound-for-pound replacement of fat, and since it is only 9-percent fat, it is possible to produce products that have 90-percent less fat than their full-fat counterparts. It contains 80-percent fewer calories per gram than fat.

In addition to the nutritional benefits, the fat replacement has added industrial benefits. First, it costs less than the food it replaces and can help manufacturers reduce material costs. Secondly, in precooked products, Nutrigras can increase moisture content, which increases product yield. For example, in research on cooked beef, the Nutrigras-enhanced product shrank 10-percent less than the beef that had not received the additive.

It is healthy, has wide-spread applicability, and is more cost effective than using full-fat products; but really, how does it taste? That is the big breakthrough. With Nutrigras, the finished product is more moist and tender. Quantitative consumer testing conducted by the company indicated that a beef patty made with Nutrigras was actually preferred to the full-fat beef patty. The unique structure of Nutrigras allows for improved flavor delivery. The construction of Nutrigras is receptive to the addition of flavors that can be carried and then released in a "burst" when consumed. This can be positioned as a point of difference for food manufacturers.

The company has been working on a number of specific applications, with the primary focus on beef, pork, chicken, and turkey. Work has also been done to enhance the performance of various baked goods, ice creams, ice cream novelties, soups, sauces, and salad dressings.

Development work and testing has been completed on beef patties. In beef, optimal results have been obtained when converting 80/20 ground beef (80-percent lean meat/20-percent fat) to 80/15/5 (80-percent lean meat/15-percent Nutrigras/5-percent fat.) Product testing is currently underway on pork sausage and chicken, and one customer is currently working on a turkey enhancement.

Nutrigras can be used to add flavors to a variety of baked goods, resulting in reduced fat and calories while enhancing flavor. Moreover, preliminary research has indicated the potential for product stability benefits from Nutrigras. Baked goods are left moister, better tasting, and the resultant product contains less fat and fewer calories.

Ice cream can be made with less heavy cream by replacing a portion of the cream with Nutrigras. Overall costs are reduced (cream is more costly than Nutrigras), and the flavor profile is enhanced and improved. In addition, unique flavors can be obtained through customized formulations. Nutrigras can act as a stabilizer and reduce the use of extraneous gums and emulsifiers that are expensive and clutter product labels. Nutrigras has also demonstrated the ability to reduce the negative freeze/thaw characteristics of conventional ice cream.

Additionally, soups can be flavor enhanced, better tasting, and have improved mouth feel. Low-fat sauces and salad dressings can be improved in similar fashions.

The company has been able to repay the help provided by NASA by contributing to the Space Agency's astronaut diet. The Nutrigras fat substitute can be used as a flavor enhancer and shelf-life extender for use on the ISS. ❖

Nutrigras™ is a trademark of H.F. Food Technologies Inc.

Additive Transforms Paint into Insulation

Originating Technology/NASA Contribution

The heat generated by wind resistance and engine exhaust during the launch of a space shuttle is potentially damaging to the casings on the solid rocket boosters, which provide over two-thirds of the initial thrust needed to propel the spacecraft into orbit. To protect this important equipment, in the 1980s, engineers at Marshall Space Flight Center developed a spray-on insulating process that was applied to the boosters' forward assembly, systems tunnel covers, and aft skirt. The process involved mixing nine chemicals into an adhesive, and then, acting quickly during a 5-hour window, applying the material. The materials were costly, and if the application was interrupted or not completed within the 5-hour window, the batch was lost. In addition to this drawback, the strength of the material was difficult to regulate, so it often chipped off during flight and splashdown, when the reusable boosters are dropped into the sea. Adding to the downside, two of the nine ingredients were harmful to the environment.

Through a Space Act Agreement in 1993, Marshall partnered with the United Technologies subsidiary, USBI, of Huntsville, Alabama, to develop an alternative to the old insulating spray. Using Marshall-developed convergent spray technology, they atomized epoxy and different filler materials to create an environmentally friendly ablative insulation material. The material, Marshall Convergent Coating-1 (MCC-1) consisted of 8-percent hollow spherical glass, 9-percent cork, and 83-percent epoxy, materials that were mixed at the time of application, at the point of release from a spray gun, which eliminated the problem of batches being ruined from interruptions and delays. The insulating paint was first flight tested in 1996 on the STS-79 mission, and was so successful that it has been employed on all subsequent shuttle flights, with virtually no observed missing or chipped paint on the spent boosters during post-flight inspections.

Partnership

David Page, founder of Tech Traders Inc., of Merritt Island, Florida, wanted assistance developing coatings and paints that create a useful thermal reflectance. After reading an article in the local paper advertising NASA technical assistance available to small business, he

contacted Marshall's Technology Utilization Office, now the Technology Transfer program office, a division of that Center's Innovative Partnerships Program.

The folks at Marshall directed Page to Kennedy Space Center, where he teamed up with a group of engineers from USBI who were developing a roof coating out of recycled car tires that would be applied using the Marshall-developed convergent spray technology. The hardest problem they faced was creating a low-cost, yet highly effective, product that was safe and non-toxic. Page had access to published NASA information regarding the heat-reflecting tile used on the space shuttle. He learned that the coating on the tile does 98 percent of the work. It appeared that if he was able to incorporate this composition into the paint, then he had a solution that would be safe, economical, and effective.

Page had an open line of communication between the engineers at Marshall and USBI. After a year of collaboration as well as additional testing with Dr. Heinz Poppendiek of the San Diego-based Geoscience Ltd., a research and development firm specializing in heat transfer, fluid flow, mass transfer, micro meteorology, biophysics, engineering design, system fabrication, product evaluation, and the measurement of thermal, mechanical, and fluid properties, Page felt that he had a product ready for market.

Product Outcome

Insuladd is a powder additive that can be mixed into ordinary interior or exterior paint, making that paint act like a layer of insulation. The company recommends two coats for optimal protection. The material is also available in a pre-mixed form.

The secret behind Insuladd is the unique propriety process that applies a coating to the microscopic inert gas-filled ceramic microspheres that make up Insuladd. When the paint dries, these form the radiant heat barrier,

Insuladd is an insulating paint additive that can be mixed with all interior and exterior paint.

turning the ordinary house paint into heat-reflecting thermal paint. The insulating materials reduce heat transfer by reflecting heat away from the painted surface by forming a heat-blocking radiant barrier on the surface that is painted.

According to Tech Traders, the product works with all types of paints and coatings and will not change the coverage rate, application, or adhesion of the paint. It can be used on walls, roofs, ceilings, air-conditioning ducts, steam pipes and fittings, and is particularly well-suited for use on metal buildings, cold storage facilities such as walk-in coolers and freezers, and mobile or modular homes.

In addition to the target market of residential and commercial buildings, customers have found a variety of other useful applications for this insulating additive. For example, Purina Feeds uses the Insuladd E-Coat, an insulating wall paint and roof paint that is a 100-percent acrylic product containing the ceramic paint additive, to paint feed storage silos to help prevent feed spoilage. The poultry industry uses Insuladd to reduce the summer heat and winter cold effects on the climate of hatcheries. Samsung uses the ceramic paint additive on military vehicles, and Hyundai Corporation's shipbuilding division uses Insuladd on its ships.

Tech Traders has continued its connection to the aerospace community by recently providing Lockheed Martin Corporation with one of its thermal products for use on the F-22 Raptor. The designers of the high-tech fighter jet were able to use one of the insulating paints on the outside of an electrical switch box that was overheating due to outside heat sources. ❖

Insuladd® is a registered trademark of The Insuladd Company.

According to Tech Traders Inc., its insulating paint additive can make homes, businesses, warehouses, ships, and other structures more energy efficient.

The insulating materials reduce heat transfer by reflecting heat away from the painted surface by forming a heat-blocking radiant barrier on the surface.

New Lubricants Protect Machines and the Environment

Originating Technology/NASA Contribution

The Mobile Launcher Platform at NASA's Kennedy Space Center is a two-story steel structure that provides a transportable launch base for the space shuttle. The main body of the platform is 160 feet long, 135 feet wide, and 25 feet high. When completely unloaded, the platform weighs about 8 million pounds. When it is carrying the weight of an unfueled space shuttle, it weighs about 11 million pounds.

To transport a fully assembled space shuttle and the Mobile Launcher Platform from the Vehicle Assembly Building to the launch pad, NASA uses a vehicle it calls a crawler. The crawler is 131 feet long, 114 feet wide, and 20 feet high (about the size of a baseball diamond), and features eight tracks fitted with 7.5- by 1.5-foot shoes that help roll the massive vehicle and its payload along.

Back in 1994, NASA sought a new type of lubricant that would be safe for the environment and would help "grease the wheels" by making the meticulous 1 mile per hour, 3-mile trek of the shuttle-bearing launcher platform to the launch pad an easier process. To satisfy the environmental requirement, the lubricant had to be biodegradable. This was especially important, since Kennedy is a wildlife refuge. To account for the size and the weight of the space shuttle/platform combination, as well as the tortoise-like pace and the distance being traveled, the lubricant had to sustain a long operating life while in use. In addition, it had to provide complete protection from the corrosive sand and the heat that are a part of everyday life at Kennedy.

Partnership

With the help of Lockheed Martin Space Operations—the contractor for launch operations at Kennedy—and private industry, the Space Agency realized that a new kind of lube could go a long way to protect the environment as well as the integrity of a space shuttle mission.

To develop a special lubricant that could meet the stringent requirements for shuttle transport, NASA and Lockheed Martin Space Operations looked to Sun Coast Chemicals of Daytona Inc. (now known as The X-1R Corporation). Founded in 1989, Sun Coast Chemicals had established a shining reputation amongst the racing circuit for manufacturing effective lubricants that were helping drivers and pit crews overcome engine and transmission problems related to heat and wear damage.

Lockheed Martin Space Operations asked Sun Coast Chemicals to formulate an advanced, environmentally friendly spray lubricant to replace the standard lubricant used during transport, and the company accepted the challenge. It brought a team of researchers, consultants, and production personnel to Kennedy to discuss a solution with NASA and Lockheed Martin personnel.

In a matter of weeks, Sun Coast Chemicals produced the solution. This new biodegradable, high-performance

When they were built, Kennedy Space Center's crawlers were the largest tracked vehicles ever made. The two crawlers, previously used to move the assembled Apollo/Saturn spacecraft from the Vehicle Assembly Building to the launch pad, are now used for transporting the space shuttles.

lubricant, coined the X-1R Crawler Track Lube, first succeeded in trial tests and then succeeded when applied directly to the crawler.

Product Outcome

In 1996, the company determined there was a market for this new development. During this time, it introduced three products that were derivatives of the base formulation it developed for the NASA application: Train Track Lubricant, which was used to solve wear problems for the Florida Power Corporation's railroad system; Penetrating Spray Lubricant, which has been applied for rust prevention, loosening corroded bolts, and lubricating joints and hinges; and Biodegradable Hydraulic Fluid, which has an oxidation life of 10,000 hours and has been used widely in processing plants, as well as in sugar, pulp and paper, marine, mining, sawmill, and heavy construction industries (*Spinoff* 1996).

Sun Coast Chemicals hit the ground running, quickly adding a gun lubricant/cleaner and a fishing rod and reel lubricant to its environmentally friendly product portfolio (*Spinoff* 1997). Now, a decade later, it has brought brand new NASA offshoot products to the market under its brand new company name.

The X-1R Corporation, of Daytona, Florida, has folded the high-performance, environmentally safe benefits into a full line of standard automotive and specially formulated racing products. At the top of this line is the X-1R Engine Treatment Concentrate, a formula that treats engine cylinder walls, bearings, cams, rings, and valve guides. It creates a molecular bond with ferrous metal, which leads to a dramatic reduction in friction and wear. It also protects against the harsh metal-to-metal contact that commonly occurs during cold starts. Other benefits of this product are increased engine life and horsepower, improved fuel economy, and reduced engine noise and operating temperatures.

The company's X-1R Plum Crazy grease is a long-life, multipurpose grease with exceptional water resistance. It is durable enough for both load-bearing and high-speed automotive applications across a wide temperature range (-30 °F to 500 °F), and is, according to the company, a leading product in suspension and open and closed bearing protection, based on extensive testing against severe heat cycles, dirt, sand, mud, and water. Other benefits include rust protection and reduced wear, drag, and friction, all leading to reduced downtime and an increase in the life of wheel bearings and other automotive parts. Plum Crazy is considered a calcium sulfonate complex

The X-1R lubricants developed originally for use on the crawlers now provide consumers with superior lubricating qualities in environmentally safe, long-lasting products.

grease, a formulation recognized by the National Grease Lubrication Institute as "an excellent technology for grease applications where heat, water, and high- or shock-loads exist."

Formulated for diesel-powered vehicles, X-1R Diesel Fuel Concentrate with Cetane Booster contains a variety of proprietary ingredients that The X-1R Corporation said helps: clean injectors; improve Cetane ratings; improve fuel economy; decrease black, white, and start-up smoking; reduce carbon monoxide, nitrogen oxide, and hydrocarbon emissions; eliminate engine knocking; and fight corrosion. (A Cetane rating is a measure of the combustion quality of diesel fuel, with a higher number representing improved quality and performance.)

Created specifically for racing differentials and transmissions, X-1R Synthetic Gear Fluid is super-resistant to high temperatures and contains extraordinary friction-modifying performance features, according to the company. Lower internal friction means less wear and lower operation temperatures. The end result is longer, trouble-free operation. X-1R Synthetic Gear Fluid is compatible with most manual transmissions and rear-end components, including quick change rear-ends.

Another synthetic product, the X-1R Air Tool Lube, is specially formulated for all piston-type and rotary air tools, as well as inline oil systems and any pneumatic air tool that requires an oil lubricant. It contains inhibitors that attack moisture and other contaminants that may prevent air tools from achieving maximum performance.

On the whole, the entire X-1R automotive product line has stood up to rigorous testing by groups such as the American Society of Mechanical Engineers (New York), the Department of Mechanical Engineering at Oakland University (Rochester, Michigan), Morgan-McClure Motorsports (Abingdon, Virginia), the Swedish National Testing and Research Institute (Boras, Sweden), and the National Power Corporation (Quezon City, Philippines).

In Dawsonville, Georgia, Elliott Racing is building some of the fastest racing engines in the country. Formed

Penn Synthetic Reel Oil, Precision Reel Grease, and Rod & Reel Cleaner now protect against corrosion and rust.

Penn Synthetic Reel Oil, Precision Reel Grease, and Rod & Reel Cleaner are now providing optimum lubrication for longer casting and smoother retrieval.

by brothers Dan and Ernie Elliott (Ernie is recognized for building winning engines for NASCAR's Winston Cup), the group recently used X-1R racing formula while testing an engine with a dynamometer—a machine that measures torque and rotational speed to determine engine power.

"It was unbelievable," said Dan Elliott when speaking of the results. "We've worked for 2 weeks to gain 2 horsepower, and here we just poured the X-1R in, and picked up 8 horsepower. But, besides the horsepower, you are picking up fuel mileage, and that plays such a critical role also, because races are won and lost on fuel mileage. What a bonus!"

Fully aware that not everybody is an automotive mechanic or an engine builder, The X-1R Corporation markets "handy packs" for simple jobs around the house. Consisting of multipurpose, multiuse lubricant and grease, these handy packs stop squeaks, reduce friction,

protect against rust and corrosion, free up stuck parts, and repel moisture. They are ideal for doors, garage doors, locks, windows, hinges, washing machines, ceiling fans, electric shavers, exercise equipment, shop tools, and lawn equipment, among many other items.

In 2003, The X-1R Corporation teamed up with Philadelphia-based PENN Tackle Manufacturing Company, a leading manufacturer of fishing tackle since 1932, to jointly develop and market a line of advanced lubrication products for saltwater and freshwater anglers, under the familiar PENN name. PENN Precision Reel Grease, Synthetic Reel Oil, and Rod & Reel Cleaner are now providing optimum lubrication for longer casting and smoother retrieval, plus protection against corrosion and rust.

"The challenge was formulating the products for the marine environment, for both fresh and saltwater applications," said Edwin "E.T." Longo, executive director of special projects development at The X-1R Corporation. "Most marine equipment is subjected to a variety of extreme conditions, such as prolonged exposure to the sun and the corrosive nature of salt and water. The lubricants we developed for NASA had to meet the same stringent demands, so we were able to develop products for PENN that are far superior to anything on the market today."

Following rigorous testing over a 2-year span, PENN committed to using the specially designed X-1R grease as the PENN Precision Reel Grease in early 2006. It is now applied to all reels leaving the company's factory.

The decision came after extensive research to identify the best possible option. "We removed the plate, which houses the gears and bearings, from a number of 965 International Baitcaster reels. A team of engineers cleaned them and coated each one with a different grease out of a selection of five well-known potentials, then we put them through a torture test," said Brent Kane, PENN's national sales manager.

"Torture included repeated dunk-and-dry cycles in saltwater baths, saltwater spray tests, and long days on

the roof in the punishing summer sun," Kane continued. "We wanted to put an accelerated life span on the reels."

According to Kane, the test subject coated with the X-1R formulation that would eventually become the PENN Precision Reel Grease was the only one to withstand the test. He noted that evaporated water left salt crystals emulsifying in the other test subjects that led to "extreme, holes-eaten-through-metal corrosion." As for the X-1R-coated reel plate, Kane noted it operated just like new. "The pinion dropped in and out, and it engaged perfectly. X-1R's PENN Precision Reel Grease displaces water and will not emulsify with saltwater, so there were no corrosion problems after getting it submerged," he concluded.

The Precision Reel Grease, Synthetic Reel Oil, and Rod & Reel Cleaner are part of the co-branded label belonging to PENN and The X-1R Corporation, and are available at a wide array of local bait and tackle shops and at large outdoor equipment retail chains, including Wal-Mart, Cabela's, Sports Authority, Dick's Sporting Goods, Sport Chalet, Big 5 Sporting Goods, Modell's Sporting Goods, Gart Sports, Academy Sports + Outdoors, Gander Mountain, G.I. Joe's Inc., Boater's World, Bass Pro Shops, and West Marine. The items are also available through the Army and Air Force Exchange Service, which provides merchandise and services to active duty, guard, and reserve members, military retirees, and family members.

These three fishing products, as well as all other X-1R advanced lubricants manufactured by The X-1R Corporation, have been officially recognized by the Space Certification Program, which is managed by the Space Foundation in cooperation with NASA. The company was also honored as the 34th inductee into the Space Foundation's Space Technology Hall of Fame in 2000. ❖

X-1R® is a registered trademark of The X-1R Corporation.

Environmental and Agricultural Resources

NASA's research helps sustain the Earth and its resources. The technologies featured in this section:

- Map, monitor, and manage Earth's resources
- Provide environmental data
- Save energy and prolong motor life
- Prevent corrosion in steel and concrete structures

Advanced Systems Map, Monitor, and Manage Earth's Resources

Originating Technology/NASA Contribution

A "revolution in remote sensing" took place in the mid-1980s, when Dr. Alexander F.H. Goetz and his colleagues at the Jet Propulsion Laboratory developed a powerful instrument called AVIRIS (Airborne Visible InfraRed Imaging Spectrometer), according to Dr. Nicholas Short, author of NASA's online Remote Sensing Tutorial. AVIRIS extended the capabilities of ground-based spectrometers, enabling the spectrum-detecting instruments to be used in the air on moving platforms.

In the early era of remote sensing, limitations in technology prevented spectrometers from being used on moving platforms mounted on aircraft and spacecraft. Essentially, the high speeds of a moving vehicle did not allow spectrometers sufficient time to accurately focus on sample features of the Earth or atmospheric targets (water vapor, cloud properties, aerosols, and absorbing gasses). All of this changed with AVIRIS. The airborne spectrometer helped to open the door for a remote sensing imaging method known as hyperspectral imaging, according to Short.

Hyperspectral imaging yields continuous spectral signatures that are captured in high-spectral resolution; this surpasses multispectral imaging methods that collect data at slower rates and in low-spectral resolution. The continuous spectral signatures, or spectral "curves," measure reflectance from the ground, water, or the atmosphere, in the wavelength region responding to solar illumination. The method is especially useful for classifying material types on the Earth's surface in fine detail, such as rock-forming minerals, soil, vegetation, and water.

Two decades after its development, AVIRIS continues to fly on aircraft today. The airborne instrument is identifying, measuring, and monitoring constituents of the Earth's surface and atmosphere in order to facilitate advancements in the fields of oceanography, limnology (the study of lakes, ponds, and streams), snow hydrology, environmental science, geology, volcanology,

The SpecTIR VNIR sensor system is a compact, commercial-off-the-shelf system suitable for use on light and unmanned aircraft. It has a spatial resolution from 0.5 to 5 meters, with a spectral range of approximately 450 to 1,000 nanometers.

soil and land management, atmospheric and aerosol studies, and agriculture.

Meanwhile, even higher above the Earth, NASA also has a hyperspectral instrument called Hyperion onboard the Earth Observing-1 satellite. Launched in 2000, Hyperion is providing a whole new class of Earth-observation data for improved Earth surface characterization.

Beginning with AVIRIS and continuing with Hyperion, hyperspectral imagery is helping to broaden NASA's understanding of the natural and man-made influences that contribute to the ever-changing Earth.

Partnership

A service-disabled veteran-owned small business concern, SpecTIR LLC is recognized for innovative sensor design, on-demand hyperspectral data collection, and image-generating products for business, academia, and national and international governments. William Bernard, SpecTIR's vice president of business development, has brought a wealth of NASA-related research experience to the company in the past few years.

Prior to joining SpecTIR, Bernard was the principal investigator on a NASA-sponsored hyperspectral crop-imaging project. This project, made possible through a **Small Business Technology Transfer (STTR)** contract with Goddard Space Flight Center, aimed to enhance airborne hyperspectral sensing and ground-truthing means for crop inspection in the Mid-Atlantic region of the United States. With Goddard's support and access to a wealth of the Center's hyperspectral resources, Bernard and his research team established a program to collect crop imagery from an aerial-mounted hyperspectral sensor and to correlate this imagery with ground-truthed field data to further define crop classifications and create spectral signature data libraries.

Bernard sought to make the spectral libraries commercially available as a means of assessing crop problems in near real time. According to his initial proposal to Goddard, early detection of crop problems directly affects the cost of treatment as well as crop yields in a season. Furthermore, the proposal stated that the data resulting from this program could help alleviate both the environmental and economic issues tied to the use of pesticides, herbicides, and fertilizer in proximity to the Chesapeake Bay watershed, which stretches across six Mid-Atlantic States, including Maryland's Eastern Shore, where Bernard's research took place.

Under the NASA-supported program, Bernard collected several seasons' worth of crop data from a DuPont agricultural research farm in Delaware and from an oil spill in tidal wetlands in Virginia. These efforts lasted several years and resulted in progress toward commercial applications for hyperspectral data.

Today, Bernard and this team of researchers have applied the knowledge and techniques generated from their NASA-supported hyperspectral research to commercial agricultural/environmental products and applications with SpecTIR.

Product Outcome

Headquartered in Reno, Nevada, with additional offices in Easton, Maryland; Manassas, Virginia; and

Hyperspectral data has many uses in agricultural applications, including for precision farming, irrigation, assessing crop health, disease monitoring, and determining soil diversity.

Santa Barbara, California, SpecTIR has carved out many niche markets for its airborne hyperspectral data products. Areas of application include precision farming and irrigation; oil, gas, and mineral exploration; pollution and contamination monitoring; wetland and forestry characterization; water quality assessment; and submerged aquatic vegetation mapping.

According to the company, many of its clients are environmental management firms and agricultural groups that require broad-location surveys and base mapping that cannot be characterized by traditional mapping technologies. In environmental monitoring, hyperspectral analysis can uncover detrimental soil erosion, as well as natural oil seeps and unnatural, man-made oil spills that are polluting natural resources. In farming, the application of water, pesticides, or fertilizer can be tailored to the needs of crops, based on conditions exposed in hyperspectral imagery.

SpecTIR's "HyperSpecTIR" suite of hyperspectral instruments incorporate visible and near infrared (VNIR) and short wavelength infrared (SWIR) sensors to capture broad, detailed imagery. The benefit of the dual-sensor approach includes a greater insight into the spectral and spatial content of a scene that is often not possible from single sources of imagery, according to SpecTIR. In agriculture, for instance, any two plants can appear similar to the casual observer, but hyperspectral imagery can reveal important spectral and spatial differences that other image sources cannot, such as chemical composition. The spatial accuracy of the imagery products conforms to accepted international mapping standards and can be incorporated into commonly used analysis packages.

The SpecTIR VNIR sensor system has a spatial resolution from 0.5 to 5 meters, with a spectral range of approximately 450 to 1,000 nanometers. It is a commercial-off-the-shelf system that can be combined with ground-truthing mechanisms or other data sets to meet client-specified classifications. The compact size of the VNIR system makes it suitable for use on light and unmanned aircraft.

The SpecTIR SWIR sensor system has a spatial resolution of 1.31 meters at 1,000 meters of altitude and a spectral range of 970 to 2,450 nanometers. The spectral resolution and number of spectral channels is user-selectable during collections with a maximum number of 254 spectral wavelength bands. (The high number of bands allows for detection of minute variations in spectral signatures.)

Today, SpecTIR has brought its NASA connection full circle, as it continues to maintain a relationship with Goddard through programs at the University of Maryland in College Park, Maryland, and at the U.S. Department of Agriculture campus in Beltsville, Maryland. Additionally, work continues on the integration of hyperspectral data with laser imaging detection and ranging systems and other commercial-off-the-shelf technologies. ❖

Sensor Network Provides Environmental Data

Originating Technology/NASA Contribution

The National Biocomputation Center is a joint partnership between the Stanford University School of Medicine's Department of Surgery and NASA's Ames Research Center. Founded in 1997, the goal of the Biocomputation Center has been to develop advanced technologies for medicine. Researchers at this center apply 3-D imaging and visualization technologies for biomedical and educational purposes, as well as support NASA's mission for human exploration and development of space. It is the test bed for much of NASA's advanced telemedicine research.

Telemedicine, the remote delivery of medical care, is important to the Space Agency, because often times those who are in need of clinical care and monitoring are as far away as the International Space Station (ISS), orbiting roughly 240 miles above the Earth. Researchers on the ISS often have backgrounds in aeronautics, physics, geology, and engineering, and are expected to conduct a wide variety of experiments in these fields, as well as perform sophisticated repairs and construction projects. While these astronauts are always well skilled and capable, what they are usually not, are medical doctors. Even if a crewmember were a doctor, though, the strict equipment weight restrictions and tight quarters aboard the orbiting laboratory would prevent the station from supporting a clinic full of medical testing equipment. The approach of telemedicine, then, aims to give astronauts in space access to a full range of medical expertise and tests, while leaving bulky equipment and large medical staffs on the ground.

With astronaut health being a top priority of the Space Agency, astronauts manning the ISS year-round, plans to set up a permanent research station on the Moon (about 238,900 miles away), and eventual travel to Mars (a whopping 46,500,000 miles away), it becomes clear that NASA has a great deal invested in learning how to monitor astronaut health and provide emergency care, while keeping the medical support facilities and crews on Earth.

These techniques and technologies developed for space travel also have applications here on Earth, where some areas are so remote that they may seem as easily accessible as the Moon.

Toward these efforts, a team of researchers at the joint research center developed a personal physiological monitoring device called Lifeguard. The device is an unobtrusive, easy-to-wear system of lightweight, rugged medical sensors. It is capable of logging physiological data, as well as wirelessly transmitting it to a portable base station computer for display purposes or further processing. The system was extensively tested for

monitoring people in remote locations performing high-risk activities, including mountain climbers, and astronauts training underwater at the NASA Extreme Environment Mission Operations facility in Florida. The system proved successful and generated a great deal of interest in the medical community, for athletic training, for first responders, and for military field use. Interestingly, though, the first commercial application of the technology is for environmental monitoring.

Partnership

In early 2005, researchers at the National Biocomputation Center formed a spinoff company, Intelesense Technologies, to provide integrated global monitoring systems, using the sensors developed at the center for monitoring astronaut health in space. One of nine companies to spin off from work conducted at the center, Intelesense uses the monitoring systems to help researchers understand how environments and people are linked, in order to monitor and protect natural resources, predict and adapt to environmental changes, and provide for sustainable development, as well as to reduce the costs and impacts of natural disasters and provide an effective and intelligent response to such disasters.

Dr. Kevin Montgomery, technical director at the center and chief executive officer of Intelesense, has a history of developing systems for image processing, 3-D reconstruction, visualization, and simulation of biomedical imaging data for space-related research at Ames. On the company's team, Dr. Carsten Mundt, the chief technology officer, is also actively engaged in several NASA-related studies, mostly in developing vital sign monitoring systems for astronauts and microsatellites. Montgomery has yet another NASA connection on his Intelesense team in Valerie Barker, who worked with

Intelesense Technologies provides global integrated monitoring products and services for environmental, public health, and other data.

Ames in 2002, designing free-flyer satellites for biological research in space.

The company's corporate offices are in Honolulu, and it has research and development offices in Milpitas, California, as well as field offices with collaborative partners in deployment zones worldwide. Montgomery, who also holds a position as an adjunct associate professor at the University of Hawaii's John A. Burns School of Medicine, where he works with other researchers developing surgical simulators, was visiting the island university when the idea for the company gelled.

He was visiting the Hawaiian school and met with Dr. Kenneth Kaneshiro and Michael Kido of the Center for Conservation Research and Training, a research program within the Pacific Biosciences Research Center at the University of Hawaii at Manoa that was established to address the rapid extinction of various species unique to the islands. While explaining the work done in developing Lifeguard, Montgomery had the realization that the technology could also be used for environmental monitoring. Kaneshiro and Kido invited him to a workshop on environmental sensing, at which he had a remarkable experience.

Participants of the workshop were flown by helicopter to what Montgomery describes as "one of the most pristine, unspoiled, biodiverse, untraveled, remote parts of the island of Kauai, where only a handful of humans have been over the past hundred years.

"After the helicopter took off," he recalls, "leaving us all there, the indescribable uniqueness impressed upon me the importance of preserving and protecting places like these."

He took this newfound excitement back to the NASA/Stanford lab, where he insisted that Mundt accompany him on the next trip back. Both researchers were impressed with the magnificence of the unspoiled area and agreed that the wireless sensor networks could be used to help in its conservation.

As Montgomery explains, "The need exists everywhere to understand the interrelationships of humans with their environment and, in order to do that, we need to acquire and integrate information from many sources and visualize and understand it in intuitive ways—that's what Intelesense is all about."

Product Outcome

Employing networks of wireless sensors for air, water, weather, and imagery, and then integrating the sensor information with other data sources, Intelesense helps clients better understand interrelationships in a wide variety of areas, including environmental preservation, monitoring waterborne illnesses, detecting infectious diseases, and providing remote health care. Current projects range from protecting the environment, to tracking emerging infectious diseases like avian influenza (bird flu), and to helping people from around the world connect and interact with each other to better understand their environment and themselves.

The company does this by deploying a worldwide wireless sensor network that communicates data from anywhere in the world, and then integrates with data from other sources automatically. The aggregated data is then turned into real-time advanced, informative graphical displays.

The company developed three technologies to accomplish the collection, sharing, and presentation of the aggregated data.

The first component is the InteleCell, a small, remotely deployable, wireless, secure data acquisition platform. Global Positioning System (GPS)-enabled, the rugged device interfaces with a variety of sensor types, such as those that monitor water, weather, air, and soil, as well as imagery and biosensors. The ultra-low-power device is rugged enough to be left unattended in harsh regions for data collection, but it is also small enough to be used as a hand-held geographic information system device or data logger. Running off of a battery sustained by solar power

Using NASA-developed technology, Intelesense makes rugged, wireless sensor devices for air, water, weather, and imagery that communicate their data over the Internet from anywhere in the world, integrate with data from many other sources automatically, and provide real-time advanced analysis and intelligent, collaborative visualization.

and several sleep modes, the remotely programmable InteleCell addresses the challenges of deploying real-world, long-range, unattended networks in areas that are often difficult to access. It was designed by Mundt and Barker and is a direct result of their NASA work on microsatellites. In fact, according to Montgomery, "The InteleCell is essentially a microsatellite on the ground—with sensors, radio telemetry, and self-powered."

Multiple InteleCells form the second component, a self-organizing InteleNet, a global, wireless sensor network that provides users with advanced graphics and analysis capabilities. The InteleNet is capable of collecting data from many sensors and sending this information to users over the Internet, providing real-time access to sensor data. The data can then be integrated with information from a variety of other sources, such as public health records or historical climate data, providing users with a more complete picture of the area being analyzed.

The third technology, InteleView, is an interactive 3-D program that allows the user to access the secure, real-time data from around the world. Built upon the NASA World Wind software platform, it provides an intuitive interface that helps users navigate multiple, diverse data sets and display the data in meaningful ways.

One innovation from Intelesense Technologies is a network of integrated water, weather, and other sensors with GPS localization over a custom wireless voice/data network with secure uplink to an Internet-based GIS Web site.

Altogether, the three-component system can be used for a variety of applications, including, but not limited to: weather monitoring, water and air quality monitoring, biotelemetry, image and audio capture, field data collection, earthquake monitoring, tsunami warning, buoy networks, emergency telemedicine, and disaster response.

With the company headquarters in Honolulu, much of its initial testing of the network technology took place in that region, on the island of Kauai. An average rainfall of over 460 inches per year makes Kauai one of the wettest places on the planet. This rainfall rushes through steep-walled gorges from 5,148 feet at the top of the island's extinct volcano, Mount Waialeale (Hawaiian for "rippling waters") to the sea far below. This area, with its lush vegetation and cascading waterfalls, is as beautiful as it is inaccessible. With no cellular coverage, and much of the area only accessible via helicopter, it proved the perfect test bed for the radio frequency-based, self-powered electronic sensors dependent upon line-of-sight transmission and solar power.

The Kauai test zone was in the Limahuli Valley, which, according to the ancient Hawaiian method of land division, consists of four ahupua'a: Waipi'o, Lumahai, Wainiha, and Haena (an ahupua'a is a strip of land that extends from a mountain to the sea). Inhabitants of these areas lived off of the land and developed a deep-felt appreciation for it, a holistic understanding of the interconnectedness of terrestrial, freshwater, and marine ecosystems that supported them. These well-preserved tracts were the ideal location to test the network's use as an environmental monitoring system, not just as extreme, worst-case scenarios of terrain and climate, but also because they needed to be monitored and maintained in their pure states.

The network was comprised of a series of InteleCells: sensor computers connected to a radio frequency module, a large battery and enclosed in a ruggedized case, with strong, weatherproof external connectors for the antennas, solar panels, and sensor. The units were strategically placed throughout the canyons, then left to gather their data and send it to the central base station, via repeaters, where the secure data were then sent to a central server over the Internet. Again, the area had to be accessed by helicopter, as the canyon walls are too steep for hiking to be a viable option. The helicopter hovered at a ridge top, and the researchers had to jump out in order to gain access to this ultra-remote area.

The sensors were deployed to monitor water quality parameters in the streams, stream depth and flow, weather, and rainfall, among other factors. The sensors also provided live video and photographic imagery. Despite the obstacles presented by the harsh environment, the sensor network was a success, and it is still in place today. In fact, the company has even added a prototype system for tracking two feral goats wearing GPS collars.

Intelesense has since spread its operation to other parts of Kauai, recently establishing a site as part of a conservation and educational outreach project in collaboration with the National Tropical Botanical Garden at Lawai.

Elsewhere in Hawaii, another network has been established for use by Maui Land and Pineapple Company Inc., for use in agricultural monitoring and preservation in a managed land region. On Oahu, the company has established a small urban test site at the

University of Hawaii campus in Manoa for research and educational purposes.

This system is, according to Jane Herrington, program director for the National Science Foundation's Office of Experimental Program to Stimulate Competitive Research, "the most advanced environmental sensing network in existence," but the time spent on the islands of Hawaii taught the founders of Intelesense more than just how to monitor and improve the environment. During the time on the islands, the company has learned to embrace a series of traditional Hawaiian values: kokua, kuleana, pono, lokahi, malama, and hiki no (respectively: helping, responsibility, doing the right thing, unity,

Intelesense Technologies produces inexpensive, real-time, field-deployable systems for environmental monitoring that integrate data from many sources (including public health data) and provide for easy analysis and intelligent, collaborative display.

taking care of the land and each other, and enthusiasm). The Intelesense team, seeing the interconnectedness between the native people and the land, realized that it is not just people affecting their environments, but the environments affecting people. With that understanding, combined with the commitment to harness their abilities for helping people connect with their environments and each other, the company steered its technology toward monitoring the interconnectivity between environments and public health.

Environmental factors and public health are often inextricably linked, and waterborne illnesses affect millions of people worldwide, especially in remote, undeveloped areas like Vietnam, where, according to the United Nations Development Program (UNDP), 80 percent of human illness in rural areas is caused by waterborne disease or pollution, and 32 million people (about 36 percent of Vietnam's population of 90 million) do not have access to clean water. According to the UNDP, worldwide, the situation is similarly appalling, with 5 million people dying every year due to waterborne illness, and waterborne illness is implicated in 60 percent of infant mortalities.

As part of a collaboration with the University of Hawaii, the U.S. Department of Health and Human Services, the U.S. Army Medical Research and Materiel Command's Telemedicine and Advanced Technology Research Center, the Vietnamese Academy of Science and Technology, and the Hanoi School of Public Health, Intelesense deployed an advanced system of environmental monitoring sensors that will be integrated with public health data to research how waterborne illnesses form.

Intelesense realized the importance of being able to link environmental factors with public health data. As Montgomery explains, "If we could show that humans and their environments are linked in this way, our system could have a big impact on the world by monitoring drinking water supplies and producing alerts, thus

potentially impacting lots of people and preventing waterborne illness."

This system, which is also tracking other major public health concerns like avian influenza, integrates data from a network of widely deployed sensors with daily public health information, providing governments, public health workers, and researchers with real-time statistics on emerging infectious diseases. This timely information would allow for rapid assessment and intervention in the event of an outbreak.

Intelesense has also set up shop in Ethiopia, one of the harshest, driest regions of the world and the third most populated nation in Africa. There, the company is developing a network for communicating public health information from 126 remote medical clinics to 5 corresponding hospitals. The sensors connect all these players with a robust, wireless infrastructure, in an area where there is no reliable cellular or telecommunications network, and even power supplies are unreliable.

The InteleCell sensors designed for this deployment interface with PDA sensors that provide patient information, interact with radio frequency identification tags on medical and blood product packages, and also provide real-time, two-way, self-powered video telecommunication. They are part of a large-scale antiretroviral (ARV) study, for which patients receiving treatment are regularly monitored at clinics, and the information is then collected, reviewed, and analyzed at the hospitals. The hospitals can use this system to track supplies of ARV drugs and ensure that sufficient supplies are in areas where the drugs are currently most needed, as well as conduct virtual training classes. With these classes, doctors at hospitals can telementor clinic staff and spread the medical expertise from hospitals to the remote villages. Like in Vietnam, this Ethiopian network will also be used to track the spread of bird flu.

Intelesense is currently planning future deployments in other areas of the Pacific, including Palau, Palmyra, and Okinawa; and in Thailand. ❖

Voltage Controller Saves Energy, Prolongs Life of Motors

Originating Technology/NASA Contribution

In the late 1970s, Frank Nola, an engineer at NASA's Marshall Space Flight Center, had an idea for reducing energy waste in small induction motors. The idea, a method to electronically adjust the voltage in accordance with the motor's load, was patented in 1984. The voltage controllers have become known as Nola devices, and they are still as useful today as they were more than 20 years ago, as they can be applied anywhere an AC induction motor is being used at a constant speed but with a variable load. These have the ability to save operators a great deal of energy when the motor is lightly loaded, which translates into savings in cost and resources.

Partnership

A recent report by the Energy Information Administration suggests that between 2003 and 2030, worldwide electricity consumption will double. Combine that with rising energy prices and concern over damage to the environment caused by energy creation, and the idea for a business that helps conserve electrical energy makes perfect sense. This was the thinking of Power Efficiency Corporation, of Las Vegas, when it licensed NASA's voltage controller technology from Marshall in 1985. The company has managed in the following years to make patented improvements to the technology and create thousands of these devices, marketing them throughout the world as the Performance Controller and the Power Efficiency energy-saving soft start.

The soft start functionality gradually introduces power to the motor, thus eliminating the harsh, violent mechanical stresses of having the device go from a dormant state to one of full activity. This prevents it from running too hot and increases the motor's lifetime.

Product Outcome

According to the company, electric motors consume about a quarter of all electricity used in the country.

Power Efficiency's energy-saving soft start employs electric circuits to monitor power use and motor workload, matching the amount of power to the workload.

The energy-saving soft start allocates power in direct proportion to the motor's required workload, eliminating wasted electricity. Power Efficiency Corporation's core technology is based on patented improvements to NASA technology.

The device can sense if the motor is lightly loaded or idling, which is when the motor is least efficient, and then ramps the power usage down while maintaining the motor at a constant speed. This feature can reduce power consumption by up to 40 percent when the motor is lightly loaded. This is a great advantage for any motor that routinely runs under variable loads.

Escalators and elevators with motor-generator (MG) sets are prime examples of the types of machinery that benefit from energy savings the device provides. Take, for example, an escalator: Sometimes it is full of people; sometimes it may have a single passenger; and other times it may be empty. The motor inside the escalator must be large enough to handle the maximum possible load—a full escalator—but that rarely happens. When the escalator has few passengers, the motor driving it is lightly loaded and wasting energy. This waste of energy is addressed by the Power Efficiency energy-saving soft start, as it senses the load and instructs the motor to use just the right amount of energy (usually much less) for the job.

The product can pay for itself through the reduction in electricity. According to Power Efficiency, the product often pays for itself within 3 years, depending on the duty cycle of the motor and the prevailing power rates. In many instances, the purchaser is eligible for utility rebates for the environmental protection it provides.

Common applications of Power Efficiency's soft start include mixers, grinders, granulators, conveyors, crushers, stamping presses, injection molders, elevators with MG sets, and escalators. The device has been retrofitted onto equipment at major department store chains, hotels, airports, universities, and for various manufacturers. The technology has been incorporated into products Power Efficiency Corporation makes on a private label basis for other manufacturers. For example, Rockford, Illinois-based, Rapid Granulator, an international manufacturer of granulators for the plastics industry, uses the device under the name Energy Pro.

Power Efficiency Corporation develops and markets advanced energy-saving technologies for electric motors, based on its licensed NASA technology. Its products include a motor efficiency controller designed to increase the efficiency of lightly loaded electric motors in escalators, elevators, grinders, granulators, mixers, saws, and stamping presses.

KONE Inc. markets the voltage controller as the EcoStart. KONE, which has its U.S. headquarters in Moline, Illinois, is one of the world's largest elevator and escalator manufacturers. KONE was founded in 1910 and has about 29,000 employees in 800 service centers in 40 countries.

Direct customers of Power Efficiency Corporation's energy-saving soft start include Seattle-Tacoma International Airport and Honolulu International Airport, as well as the Toronto and Anchorage airports. In airports, the devices often control escalators and moving walkways. The device can also be found installed at the State University of New York's Fashion Institute of Technology, The George Washington University, in Washington, DC, and in department stores like Saks Fifth Avenue and Mays.

The company has also partnered with several Las Vegas neighbors, large casinos whose escalators and elevators ferry millions of tourists per year, 24 hours per day. The company is continuing to expand into the hospitality industry in Las Vegas, a city notorious for its power usage.

Power Efficiency Corporation is also looking into developing products that make electric motors more efficient for home appliances like air conditioners, clothes dryers, and refrigerators, three of the largest electricity consumers in most homes. ❖

Energy Pro™ is a trademark of Rapid Granulator.
EcoStart™ is a trademark of KONE Inc.

Treatment Prevents Corrosion in Steel and Concrete Structures

Originating Technology/NASA Contribution

NASA's Kennedy Space Center is located on prime beachfront property along the Atlantic coast of Florida on Cape Canaveral. While beautiful, this region presents several challenges, like temperamental coastal weather, lightning storms, and salty, corrosive, sea breezes assaulting equipment and the Center's launch pads. The constant barrage of salty water subjects facility structures to a type of weathering called spalling, a common form of corrosion seen in porous building materials such as brick, natural stone, tiles, and concrete. In spalling, water carries dissolved salt through the building material, where it then crystallizes near the surface as the water evaporates. As the salt crystals expand, this creates stresses which break away chips, or spall, from the surface, causing unsightly and structural damage.

The potential for corrosion heightens as concrete structures age, because over time concrete loses its acidity, or pH. When it starts out, poured concrete has a high pH value, between 11 and 13, which helps to inhibit corrosion. Over time, though, this value drops, and when the pH value dips into the 8 to 9 range, there is potential for corrosion of the steel reinforcing bars, or rebar, causing further structural concerns.

In the mid-1990s, to protect the rebar, NASA developed an electromigration technique that sends corrosion-inhibiting ions into rebar to prevent rust, corrosion, and separation from the surrounding concrete. An ounce of prevention is worth a pound of cure, and with the help of Florida's Technological Research and Development Authority, an independent state agency that partnered with Kennedy on technology transfer initiatives, the Center began working with Surtreat Holding LLC, of Pittsburgh, Pennsylvania, a company that had developed a chemical option to fight structural corrosion in 1997. Surtreat's method was to apply its anti-corrosive solution product, TPS-II, to the surface of a corroding concrete slab, where it would then seep through to the rebar, coating it and preventing further corrosion.

Combining Surtreat's TPS-II with electromigration fit well in the Kennedy dual-use program, part of NASA's technology transfer and commercialization effort. The cooperative effort involved Surtreat providing NASA with the corrosion-inhibiting chemical and concrete test slabs, along with technical support as needed. Kennedy provided testing specifications and procedures, then prepared the concrete with the Surtreat chemical and carried out an environmental evaluation of the treatment. Kennedy's materials scientists reviewed the applicability of the chemical treatment to the electromigration process and determined that it was an effective and environmentally friendly match, suitable for use at the NASA facility.

Ten years later, NASA is still using this combined approach to fight concrete corrosion, and it has also

developed a new technology that it believes will further advance these efforts. The technology is a liquid galvanic coating, applied to the outer surface of reinforced concrete, to protect the embedded rebar from corrosion. The coating contains one of several types of metallic particles—magnesium, zinc, or indium. An electrical current established between metallic particles in the applied coating and the surface of the steel rebar produces cathodic protection of the rebar. The current forces a flow of electrons from the coating (anode) to the rebar along a separate metallic connection. This surplus of electrons at the rebar (cathode) prevents the loss of metal ions that would normally occur as part of the natural corrosion process. The technology, made of inexpensive, commercially available ingredients, can be applied to the outside surface of reinforced concrete (most rebar corrosion prevention must be applied directly to the rebar) and with a conventional brush or spray, eliminating the need for expensive, specialized labor.

This new technology is currently in use at Kennedy, but NASA also saw the immediate benefits that could be gained by transferring this technology to the private sector, where decay and corrosion of concrete structures costs billions of dollars per year.

Partnership

Ten years after its initial partnership, Surtreat has partnered with NASA again by licensing the new liquid galvanic coating technology and has already put it to use. Its first test, in early 2007, was completed at the U.S. Army Naha Port, in Okinawa, Japan, a coastal facility built during the Korean War and subject to much of the same environmental stressors as those found at Kennedy.

The Surtreat coating is environmentally friendly and lasts 10 years or more, reducing maintenance costs over the lifetime of the structure.

This parking structure was suffering from water and deicing salt infiltration, causing a number of corrosion-related problems throughout the structure. The Surtreat chemicals migrated over 2 inches below the surface, halting corrosion.

where they become part of the steel and concrete matrix indefinitely. It leaves no residues, coatings, or materials that could potentially harm humans, animals, fish, or the environment. ❖

Total Performance System™ is a trademark of Surtreat International.

Product Outcome

The NASA-developed coating may be used to prevent corrosion of steel in concrete in several applications, including highway and bridge infrastructures, piers and docks, concrete balconies and ceilings, parking garages, cooling towers, and pipelines, to name just a few.

Surtreat is the ideal partner to bring this technology to the public, as the company has a proven record of providing full-service, innovative, and technical solutions for the restoration and prevention of deterioration and corrosion in steel-reinforced concrete structures. Their Total Performance System provides diagnostic testing and site analysis to identify the scope of problems for each project, manufactures and prescribes site-specific solutions, controls material application, and verifies performance through follow-up testing and analysis.

The coating lasts 10 years or more, reducing maintenance costs over the lifetime of the structure; and testing has proven that the treatment yields reductions in rebar corrosion potential, water penetration, chemical reactivity, and water-soluble chloride, while generating increases in hardness, flexural strength, and pH levels. The treatment also provides resistance to chloride penetration and problems associated with freezing and thawing of the porous structures.

Surtreat treatments are environmentally friendly, and the company focuses on preventing and minimizing adverse environmental impacts by identifying and controlling potential environmental risks in advance. The solutions used are water-soluble and environmentally safe, and in testing have shown no effect on the turbidity, pH, or dissolved oxygen content levels in water. Surtreat's formulations bond inorganic compounds to structures,

Surtreat uses topically applied, chemically reactive, migrating formulations unique to each project.

Computer Technology

NASA's work in advanced computing has led to many innovations. The technologies featured in this section:

- Simplify analysis and design
- Translate 2-D graphics to 3-D surfaces
- Improve health and performance monitoring
- Enable smarter content management
- Validate system design

Optics Program Simplifies Analysis and Design

Originating Technology/NASA Contribution

Future spaceborne astronomy missions will require telescopes with increasingly greater power, driving the dimensions of the optics and their housing structures to significantly greater sizes.

The increased size of the structures reduces the dynamic frequencies of the optical system, to the point where disturbance frequencies and structural modes significantly interact. At the same time, the requirements on dynamic stability to achieve the required optical performance are significantly tighter than for anything that has flown

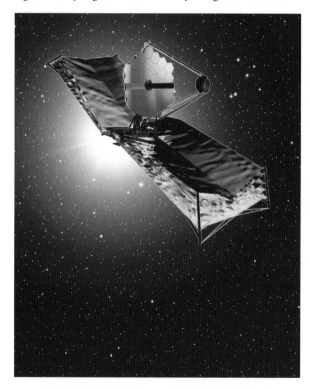

The James Webb Space Telescope is a large, infrared-optimized telescope, scheduled for launch in 2013. It will find the first galaxies that formed in the early universe, connecting the Big Bang theory to our own Milky Way galaxy.

before, and, therefore, the sensitivity to dynamic effects is correspondingly high. Finally, the physical size of the optical instruments makes fully integrated system-level testing extremely problematic. Not only are the systems too large to test in any existing environmental chambers, they are susceptible to gravity loading effects and suspension coupling that will significantly change the dynamics. Validation of the large designs must then rest on a combination of analysis and system tests.

Techniques for system tests include using Finite Elements (FE) and FE model updating tools, system identification, and using various other tools for performing optical analyses. These techniques, however, do not provide for analysis of cross-disciplinary results. Therefore, the conventional approach is to develop a requirement budget that assigns error allocations to each of the modeling teams. This approach is extremely limiting in that it does not allow requirements to be freely traded among subsystems. NASA, intent on sending larger, more powerful optics into space, resolved to find a better way to test them.

Partnership

Engineers at Goddard Space Flight Center partnered with software experts at Midé Technology Corporation, of Medford, Massachusetts, through a **Small Business Innovation Research (SBIR)** contract to design a new analysis system.

The result of the two-phase contract was the Disturbance-Optics-Controls-Structures (DOCS) Toolbox, a software suite for performing integrated modeling for multidisciplinary analysis and design. The Toolbox allows the definition of subsystem/component models, including structural models, control system models, optical sensitivities, and disturbance models. The component models are automatically coupled together to create a mathematic model of a complete physical process, using techniques that maximize the numerical conditioning, while maintaining modeling accuracy.

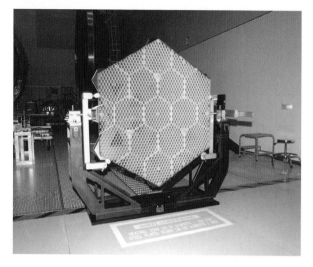

One of many segments of the mirror assembly being tested for the James Webb Space Telescope project at the X-Ray Calibration Facility at Marshall Space Flight Center. Marshall is supporting Goddard Space Flight Center in developing the telescope by taking numerous measurements to predict its future performance.

The code has been validated and applied to the following NASA astronomy projects and facilities: the Terrestrial Planet Finder Structurally Connected Interferometer Testbed (TPF-SCIT), the Terrestrial Planet Finder Coronagraph (TPF-C), the James Webb Space Telescope, and the Solar Dynamics Observatory.

Product Outcome

The purpose of the DOCS Toolbox is to integrate various discipline models into a coupled process math model that can then predict system performance as a function of subsystem design parameters. The Toolbox accepts as input the discipline models from a variety of currently available discipline modeling tools. The Toolbox then connects the discipline models and applies numerical conditioning algorithms to improve the numerical accuracy, while still maintaining model

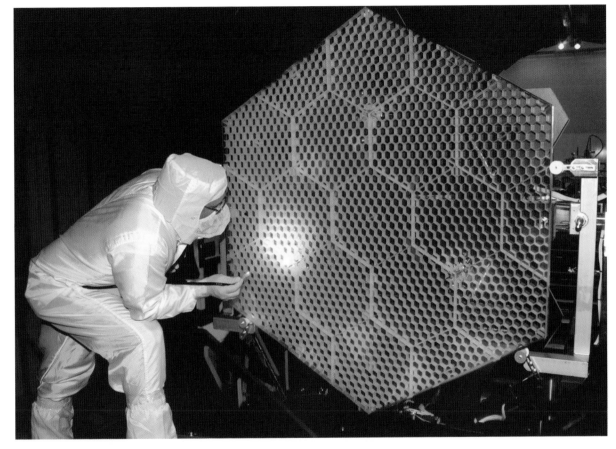

A Marshall Space Flight Center employee is inspecting one of the mirror assemblies for flaws.

methodology (a design concept that seeks to create the best-fit concept) that maps out the non-unique set of design parameters that meet requirements; uncertainty analysis for computing errors bound to performance predictions and identifying critical uncertainties; and model updating to renew component math models using measurement data.

The software has myriad benefits, including the ability to automatically couple the discipline models to create a system math model. This provides a complete description of the physical process in terms of all component design variables. The automatic coupling reduces manual effort, eliminates the chance for user error, and automatically checks unit compatibility between components.

The software tool provides a set of "canned" routines for defining parameter dependencies on typical structural parameters, as well as a set of techniques for identifying critical design parameters.

The unique formulation of the parameter-dependent math model enables the designer to formulate many problems as formal optimization analyses, where other analysis techniques would have allowed only the evaluation and comparison of a limited set of design points.

The product is being sold commercially by Nightsky Systems Inc., of Raleigh, North Carolina, a spinoff company that was formed by Midé specifically to market the DOCS Toolbox.

Commercial applications include use by any contractors developing large space-based optical systems, including Lockheed Martin Corporation, The Boeing Company, and Northrup Grumman Corporation, as well as companies providing technical audit services, like General Dynamics Corporation. ❖

DOCS® Toolbox is a registered trademark of Nightsky Systems Inc.

accuracy. It performs the analysis and redesign in a graphical framework that allows the user to define and solve the analysis problem, and it then documents results in a point-and-click environment.

The system can be optimized for performance; design parameters can be traded; parameter uncertainties can be propagated through the math model to develop error bounds on system predictions; and the model can be updated, based on component, subsystem, or system-level data.

The Toolbox allows the definition of process parameters as explicit functions of the coupled model, further enabling the exact definition of sensitivities. It also includes a number of functions that analyze the coupled system model and provide for redesign, including: critical parameter analysis that formally identifies the design variables that have the highest influence on system performance, risk, and cost; optimization of design objective functions subject to constraints on design variables; formal system trading using an isoperformance

Design Application Translates 2-D Graphics to 3-D Surfaces

Originating Technology/NASA Contribution

When it comes to solving some of NASA's most challenging technical problems, the mathematical minds that make up the Computational Sciences Branch at NASA's Glenn Research Center are ready and waiting to crunch some numbers. Calculating complex algorithms and mathematical equations like it's child's play, the group has worked out many technical issues for NASA over the years.

Bruce Auer was a member of the Computational Sciences Branch for 43 years until he retired in 2005. When he started in 1962, he worked on a wide variety of problems in the fields of aeronautics and space technology. A key component of his job was developing complex algorithms that contributed to solving mathematical problems and modeling chemical and physical phenomena. Early on, the projects ranged from solving difficult heat equations for rocket nozzle technology used to prevent nozzles from burning up with multiple restarts, to studying cutting-edge turbulent flow theories.

In the late 1980s, Glenn started developing technologies for designing and manufacturing carbon graphite composite turboprop fan blades and turbofan inlet fan blades, ranging in size from 2.5 inches to 30 inches. (Glenn specializes in the development of jet engine and turbomachinery technology.) Some of these projects required making hollow blades with internal passages to dampen jet engine noise. Several years later, in 1994, Auer's group was called on for help when one of these projects, a joint Glenn-Pratt & Whitney experimental turboprop fan blade project, was not going as expected, because the experimental blades being designed and manufactured at Glenn were failing to meet Pratt & Whitney's specifications during testing.

The technical shortcomings experienced in this joint project were traced to three-dimensional (3-D)-to-two-dimensional (2-D) commercial flattening software, which was not able to accurately flatten a 3-D surface into a 2-D surface. Flattening software of this nature is required, because the fan blade was being made from 50 to 250 separate 2-D carbon graphite composite sheets, or plies, that were cut and stacked in a mold where heat and pressure was applied to create a solid carbon graphite blade.

According to Auer, to determine any one 2-D pattern, first a 3-D surface internal to the blade had to be computed. Then, this 3-D surface had to be flattened into its corresponding 2-D pattern. When the Glenn engineers making the fan blades could not find a commercial vendor who could supply accurate 3-D-to-2-D software, they came to Auer for a solution. Specifically, the engineers asked Auer to write a 3-D-to-2-D flattening program that could accurately flatten any 3-D ply internal to a blade whose external surfaces had complex, continuously changing double curvatures.

Auer went to work on writing an optimized, 3-D-to-2-D blade-flattening software program that could address the fan blade engineers' needs. By taking several

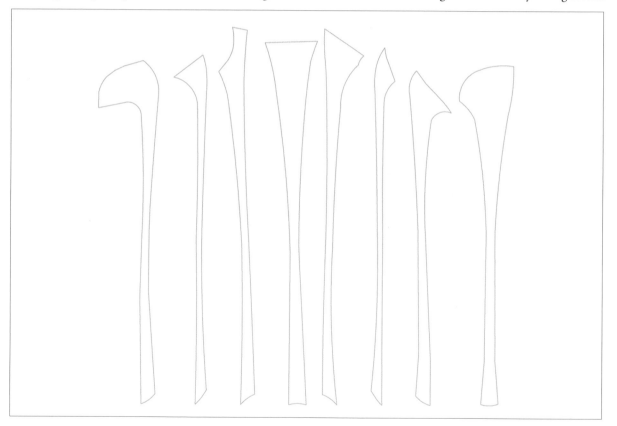

One of the technical challenges faced by the Computational Sciences Group at Glenn Research Center was creating software that was able to accurately flatten a 3-D surface into a 2-D surface.

mathematical algorithms and weaving them together with "tight logic," he came up with a solution: a program that could handle any 3-D surface fed into it.

Auer considered the outcome to be both simple and sophisticated for flattening. It could be used in either a very straightforward or advanced manner, depending on the user and his or her requirements and applications. He stated that the advantages of the program are that it is highly accurate, fast, and robust. Furthermore, he noted that it can address stretching and shrinking issues, which are inherent in some manufacturing processes.

The software program was successfully put to use in designing and manufacturing blades for the Glenn-Pratt & Whitney joint project, as well as in several other blade-manufacturing projects from other departments within Glenn over the years. All of the great results were adding up and, in 2002, several blade engineers recommended that Auer seek commercial applications for his software program.

Auer agreed with this assessment, so he pursued commercialization of his software through Glenn's Commercial Technology Office. This office accepted his proposal and arranged for his software program to be featured in the March 2002 issue of *NASA Tech Briefs*, thus giving it exposure to the outside engineering community.

Partnership

In the spring of 2005, Joseph Settipani was combing the Internet in an effort to track down a software program that could help with the flattening of fabric surfaces and, therefore, help the company cut down production time and limit the waste of materials. During his Internet search, he came across information on Auer's software for optimal flattening of fan blade patterns. After several correspondences with Auer, Settipani recognized the potential for the NASA software to bolster his Elgin, Illinois-based Fabric Images Inc. fabric-design process.

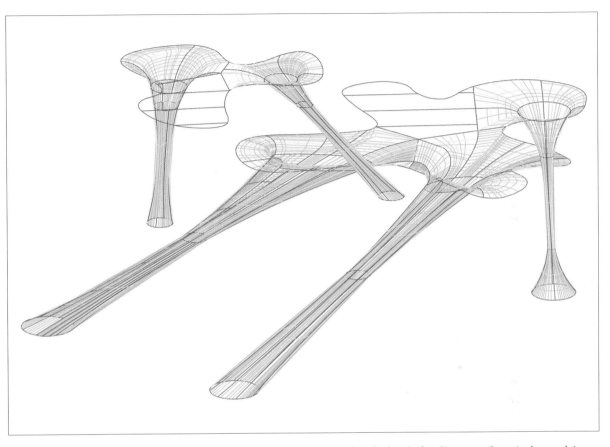

Fabric Images Inc. has the difficult task of designing the graphics for complex displays before it can see the actual, complete, 3-D displays come together.

Additional Internet research brought Settipani to the Great Lakes Industrial Technology Center (GLITeC), which works with regional companies to acquire and use NASA technology and expertise, especially from Glenn. A meeting was arranged between Settipani and Auer.

For the meeting, Auer visited the Fabric Images manufacturing plant and learned all about the company's design and manufacturing processes. Auer saw the value that his software could bring to Fabric Images and determined that the company was a great fit as a commercial partner. GLITeC subsequently facilitated an agreement that would entitle the company to utilize the NASA software code as a resource for its commercial applications.

Product Outcome

Fabric Images specializes in the printing and manufacturing of fabric tension architecture for the retail, museum, and exhibit/trade show markets. The company has the difficult task of correctly designing 2-D graphics for 3-D surfaces. Since the displays it assembles serve

Fabric Images Inc. impresses its clients with intricate and innovative displays and designs.

NASA helped overcome the unique design challenges of creating 3-D designs out of 2-D materials.

as advertisements for its clients, there is the necessity of making sure the words and logos are visually correct on the 3-D surfaces.

The goal in the design process is to achieve 2-D template-based production, based on the corresponding 3-D surface geometry. The advantages of template-based production are many, but there are many variables involved that complicate the process, as well. In order to control variables during actual production, Fabric Images

now relies on a novel and reliable flattening process based on the NASA software.

Fabric Images' fabric-flattening design process begins with the modeling of a 3-D surface based on computer-aided design specifications. The surface geometry of the 3-D model is utilized in the formation of a 2-D template, similar to NASA's flattening process. This template or pattern is then applied in the development of a 2-D graphic layout. To achieve the desired visual graphic look

on a 3-D surface, repositioning and distortion of 2-D graphics is necessary. When printed fabric graphics are not repositioned and/or not distorted for 3-D construction, fabric is wasted and money is lost in having to initiate graphic layouts and another production run.

Fabric Images developed a 2-D template pattern-based production process with the use of the NASA technology as a resource. This eliminated an entire step from the actual fabric sewing/construction process, creating an 11.5-percent time savings per project. Additional benefits included less material wasted and the ability to improve upon graphic techniques and, thus, offer new design services.

Today, Fabric Images is wowing its business-to-business partners and end-user clients with intricate and innovative displays and designs, using the NASA technology. Partners include Exhibitgroup/Giltspur (end-user client: TAC Air, a division of Truman Arnold Companies Inc.), Jack Morton Worldwide (end-user client: Nickelodeon), as well as 3D Exhibits Inc., and MG Design Associates Corp.

Although the full potential of Auer's "flattening math engine" has yet to be fully realized by Fabric Images, Marco Alvarez, the company's president, believes the future of template-based production will be bright.

"The 3-D-to-2-D templates have given us tools to take our production to a new level of efficiency. They have also opened a new avenue of business in providing graphics for difficult, amorphic shapes," stated Alvarez. ❖

Since the displays Fabric Images Inc. assembles primarily serve as advertisements for its clients, there is the necessity of making sure the words and logos are visually correct on the 3-D surfaces.

Hybrid Modeling Improves Health and Performance Monitoring

Originating Technology/NASA Contribution

Scientists and engineers have long used computers to model physical systems. Physical modeling is a major part of design and development processes, as well as failure analysis. At NASA, scientists and engineers rely heavily on physical modeling to evaluate the overall health and performance of all mission-related flight vehicles.

Hidden in the architecture of flight vehicles are computers to control and monitor their health and performance. Early versions of these computers lacked the ability to evaluate all of the operating and environmental conditions, which is important to fully understand the response of the equipment. Furthermore, traditional physical modeling methods necessitated significant computing power, making the process fairly complicated.

NASA scientists and engineers needed a simplified, but more complete, understanding of the health and performance of flight vehicles during their operation in near-real time. The comparison would allow the users to understand if the equipment was performing as expected.

Partnership

Scientific Monitoring Inc. specializes in condition monitoring and equipment health software and services for a wide range of equipment applications. NASA's Dryden Flight Research Center awarded the Scottsdale, Arizona-based company a Phase I **Small Business Innovation Research (SBIR)** contract to create a new, simplified health-monitoring approach for some of the Agency's flight vehicles and flight equipment. The main objective of the SBIR project was to create a simple software-based model that used the principles of physics, but did not require large amounts of data or computing resources to make accurate assessments in a timely fashion.

During the project, Scientific Monitoring developed a hybrid physical model concept that provided a structured approach to simplifying complex design models for use

in health monitoring. The modeling approach used a simplified analytical model and a classical data analysis technique to assess the performance of a piece of equipment and determine if it was within an expected range for the environmental and operating conditions. It allowed the output or performance of the equipment to be compared to what the design models predicted, so that deterioration or impending failure could be detected before there would be an impact on the equipment's operational capability.

The hybrid physical model was successful, as it made health and performance analyses of complicated equipment possible in near-real or real time for rapid diagnosis. Scientific Monitoring quickly realized its resulting software had broad application for many industries beyond the aerospace sector. Medical health monitoring, security, and financial transactions, for example, are other areas of application the company thought might be fitting for its new development, since each contains an underlying physical basis or characterized relationship which can be used to ensure that the behavior or performance is within expectations for the operational conditions.

Product Outcome

Marking the successful completion of the NASA-funded research project—and subsequent maturation of the model from U.S. Air Force funding—Scientific Monitoring released a commercial health- and performance-monitoring software product, based on the original modeling technology it developed for NASA. The company calls the software I-Trend for its intelligent trending, diagnostic, and prognostic capabilities.

I-Trend comes with a configurable interface, and easy-to-use back-end analysis tools to simplify Six Sigma analysis of equipment health and performance. Control charts, regression fit charts, usage charts, and many other tools provide information in an easily recognizable format.

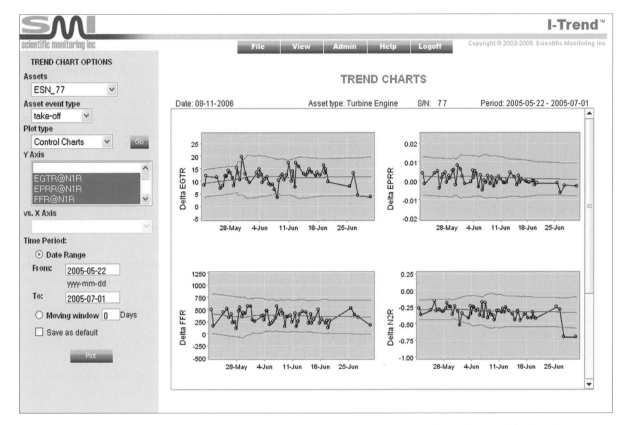

I-Trend is a Web-based condition-monitoring application specifically developed for analysis, trending, and alert functions on a wide range of equipment types.

I-Trend uses the hybrid physical model to better characterize the nature of health or performance alarms that result in "no fault found" false alarms. Additionally, the use of physical principles helps I-Trend identify problems sooner. According to Scientific Monitoring, I-Trend represents the state of the art in condition-monitoring software. It is part of the company's complete ICEMS (Intelligent Condition-based Equipment Management System) suite of software targeted for customers who have important equipment and require accurate monitoring and advanced alerting. I-Trend's advanced alerting module provides forecast capabilities that allow maintenance for many types of problems to be scheduled and planned in a proactive way, eliminating reactive behavior that occurs when alarm bells sound.

I-Trend comes with a configurable interface and easy-to-use analysis tools to simplify Six Sigma analysis of equipment health and performance. (Six Sigma is a highly structured process-improvement methodology that utilizes data and statistical analysis to gauge and enhance a company's operational performance, practices, and systems. It identifies, prevents, and eliminates defects in manufacturing and in service-related processes.) I-Trend can be used as a Web application, or specific interfaces can be tailored to allow it to operate with existing applications. It is built in a Java J2EE framework with XML (eXtensible Markup Language) interfaces to allow platform portability. Control charts, regression fit charts, usage charts, and many other charts and tools provide information in an easily understood graphical format.

I-Trend technology is currently in use in several commercial aviation programs, and the U.S. Air Force recently tapped Scientific Monitoring to develop next-generation engine health-management software for monitoring its fleet of jet engines. As for the medical, security, and financial applications that Scientific Monitoring envisioned upon completing the NASA-funded project, they just might come to fruition one day.

"The benefits of the hybrid modeling technology used in I-Trend for equipment monitoring are clear," said Dr. Link Jaw, president and chief executive officer of Scientific Monitoring. "It's the potential use in non-equipment applications which we find intriguing. It could easily evolve to allow people to monitor their own health and physical performance."

After development of the I-Trend software, the company continued the original NASA work, this time under a Phase III SBIR contract with a joint NASA-Pratt & Whitney aviation security program on propulsion-controlled aircraft under missile-damaged aircraft conditions. ❖

I-Trend™ and ICEMS™ are trademarks of Scientific Monitoring Inc.

Six Sigma®/SM is a registered trademark and service mark of Motorola Inc.

Java™ is a trademark of Sun Microsystems Inc.

Software Sharing Enables Smarter Content Management

Originating Technology/NASA Contribution

As NASA's leading organization for information sciences, the Intelligent Systems Division at Ames Research Center conducts world-class computational research to enable out-of-this-world capabilities. In particular, this division is dedicated to ushering in a new era of autonomous spacecraft and robotic exploration, as well as extending abilities in space through human-computer interactions and data analysis.

In addition to supporting NASA's missions, the Intelligent Systems Division is focused on supporting national needs by finding practical uses on Earth for its space-driven technologies. Like many of NASA's branches, it accomplishes this goal through partnerships with private industry, collaborative research initiatives with academia, and resource sharing with other U.S. Government agencies.

One of the division's latest space-based technologies that has proven practical for terrestrial use is a database storage software system called Netmark. Offering an extensible, clientless, schema-less, information-on-demand framework for managing, storing, and retrieving unstructured and/or semi-structured documents, Netmark was created to help NASA scientists query and organize complex research documents. Netmark utilizes a hybrid approach to database management by combining object-oriented data creation with relational models, using SQL (Structured Query Language) queries. This enables efficient keyword searches that cover a broad range of content, context, and relationship concepts among documents. Furthermore, it provides a single-format, virtual database view of multiple heterogeneous data sources without user-supplied database code.

All of these features make Netmark a powerful tool for managing and accessing NASA's enormous stock of complex, constantly changing unstructured and semi-structured data, as well as an open-source enterprise capable of bearing boundless commercial benefits.

Partnership

In 2004, NASA, through Ames, established a technology partnership with Xerox Corporation, in which the world's largest document management company is helping the Space Agency develop high-tech knowledge management systems, while providing new tools and applications that support the Vision for Space Exploration. In return, NASA is providing research and development assistance to Xerox so that the Stamford, Connecticut-based company can progress its product line.

"This joint venture combines the best software technology from NASA and Xerox," said G. Scott Hubbard, director of Ames from 2002 to 2006, following the announcement of the partnership. "Since both partners bring new technology to the project, we will get new tools tailored specifically for NASA needs in a very cost-effective way."

The first result of the technology partnership was a new system called the NX Knowledge Network (based on Xerox DocuShare CPX). Created specifically for NASA's purposes, this system combines the Netmark content management software created by the Intelligent Systems Division with complementary software from Xerox's global research centers and DocuShare. In its pilot stages, the NX Knowledge Network was tested at the NASA Astrobiology Institute, where researchers used it on a distributed basis across several universities to sort and analyze data. Currently, it is widely used for document management at Ames, Langley Research Center, within the Mission Operations Directorate at Johnson Space Center, and at the Jet Propulsion Laboratory, for mission-related tasks. These applications are ultimately helping NASA to manage project risk, investigate mishaps, and analyze anomalies.

Product Outcome

The research and development support Xerox has received thus far through the joint technology partnership with Ames has allowed the company to integrate the information-on-demand organizational elements of the NX Knowledge Network/Netmark software into its commercially available DocuShare enterprise content management (ECM) solution.

According to Xerox, DocuShare delivers document and content management to an organization's knowledge workers in a flexible, easily deployed, Web-based software application. The company noted that these workers are accustomed to handling vast amounts of diverse business content as part of routine operations, and are frequently asked to engage in impromptu activities to generate paper and electronic documents, as well as discussions and e-mails, from a variety of content resources. With DocuShare, the organization gives knowledge workers the necessary toolset to unlock information stored throughout the enterprise, be it in e-mail systems, filing cabinets, or network drives. Workers can then bring these diverse content types together in a searchable, indexed repository, enabling optimal information flow and rapid turnkey solutions that can translate into reduced costs and risk for the organization.

The enterprise management tool also enhances organizational collaboration and information sharing through document routing and review and Web-publishing abilities. There are also multiple levels of security built in, including password management and password rules enforcement, to protect content as necessary.

According to Xerox, the core technology simplifies deployments and reduces information technology complexity, even in the face of the typical heterogeneous mix of databases, servers, directory services, and storage systems prevalent in today's enterprises. Unlike competitive content management products, Xerox adds, DocuShare is entirely Web-based, so it works across multiple operating systems and can integrate with popular infrastructure databases and Internet browsers. Essentially, this enables DocuShare to support a mix-and-match enterprise content infrastructure and scale from small

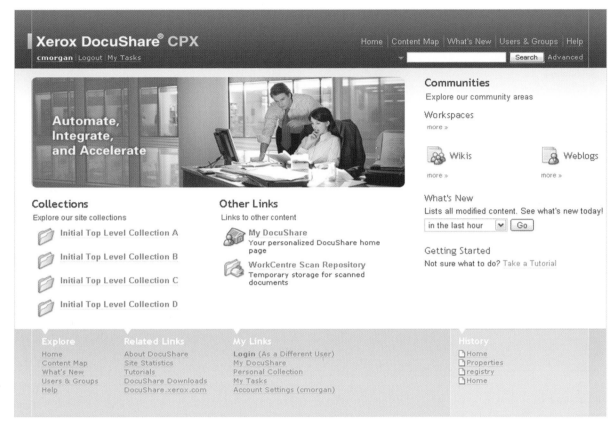

Xerox DocuShare CPX allows users to leverage the content that fuels transactional business functions—contracts, invoices, applications, and forms—more effectively and with greater flexibility than ever before.

installations to large installations (involving thousands of users) without extensive replanning or infrastructure overhauls by the customers.

In January 2006, the 4.0.1 version of the DocuShare software was named "Best Document Manager" in InfoWorld magazine's "2006 Technology of the Year Awards." Later in the year, Xerox unveiled DocuShare 5.0 and DocuShare CPX, two ECM software products built on a single technology platform. According to Xerox, organizations can use one or both applications to handle their content and document management needs, such as financial record keeping or disclosure-related government regulations.

Xerox DocuShare, now at version 6.0, was created to meet the requirement for basic, widespread content management that can scale across an organization, whereas DocuShare CPX 6.0 provides more advanced ECM features required for high-end, process-centric content applications. DocuShare CPX offers a new content assimilation and reuse feature—generated from the technology partnership with Ames—that allows XML (eXtensible Markup Language)-based content within documents to be repurposed for future use. The feature is called XDB (eXtensible Database). Xerox stated that this feature lets workers organize and manage semi-structured data, such as business forms or templates, by leveraging existing user interfaces in commonly used applications, such as Microsoft Word, Excel, and PowerPoint. With version 6.0, Xerox ships out-of-the-box applications that make the XDB content aggregation and summarization capabilities readily available in Word and Excel templates, without sophisticated programming.

"With our dual product strategy, Xerox is challenging the current enterprise content management paradigm," said David Smith, vice president and general manager of the Xerox DocuShare Business Unit. "We give customers the flexibility to mix and match both types of content management solutions on one easy-to-use platform."

This flexibility accommodates educational, financial, health care, legal service, and government organizations, among many others. In education, for example, it helps faculty and staff search and store student records, instructional guidelines, and program certifications, as well as distribute course materials for traditional and interactive e-learning initiatives. Additionally, in health care, it enables organizations to engage in document management and quickly retrieve items such as patient charts, signed consent forms, laboratory results, X-rays, caregiver notes, and clinical policies. DocuShare also provides the key capabilities for health care organizations to ensure that any potential document management solutions support compliance with procedural and regulatory standards and requirements, such as those set by The Joint Commission and the Health Insurance Portability and Accountability Act. ❖

Engineering Software Suite Validates System Design

Originating Technology/NASA Contribution

Design errors are costly. When it comes to creating complex systems for aerospace design and testing system readiness, engineering system requirements must be clearly defined, and these systems need to be tested to ensure accuracy, consistency, and safety. Testing a system, however, can require as much as 50 to 70 percent of the total design cycle time. The ability to identify potential problems early in the design cycle saves time and expense, while still ensuring safe and reliable systems. This type of research is of interest not only to the NASA Ames Research Center's Robust Software Engineering group, but to government agencies and industry, any sectors which build critical, expensive systems, such as control software for an aircraft or the U.S. Ballistic Missile Defense System's command and control system.

Partnership

To date, more than $6.5 million of government funding has been dedicated to the development of EDAptive Computing Inc.'s (ECI) EDAstar engineering software tool suite. NASA's Ames Research Center provided a significant share of this funding, through a total of five **Small Business Innovation Research (SBIR)** contracts (three Phase I contracts and two Phase II contracts). This backing from Ames allowed the Centerville, Ohio-based company to generate critical components of the software tool suite, namely Syscape and VectorGen.

Syscape is a platform-portable, customizable system design editor that utilizes a hierarchical block diagram structure, multiple design views, and user-defined plug-ins to capture executable specifications of multi-disciplinary systems. These executable specifications can be used to analyze concepts and requirements; balance risk and performance trade-offs among the various subsystems; develop system and subsystem specifications;

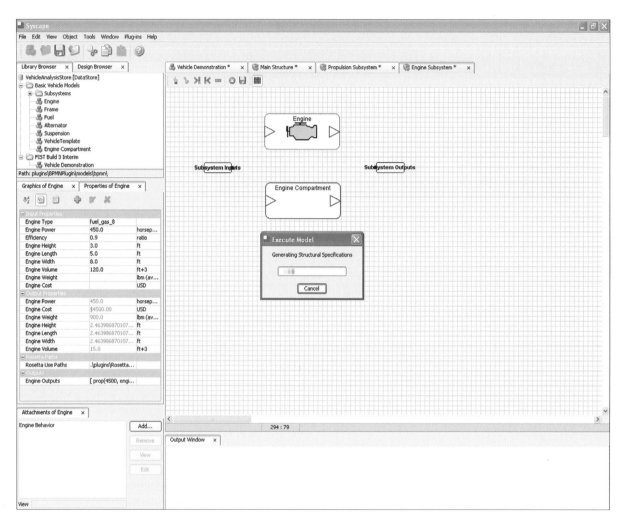

With graphical editing tools, EDAstar-based solutions can be used to rapidly create high-level, high-confidence design concepts.

and apply formal, mathematically rigorous techniques to ensure safety, accuracy, and consistency. Once created, executable specifications can be used in conjunction with VectorGen to automatically generate tests to ensure system implementations meet specifications. According to the company, the VectorGen tests considerably reduce the time and effort required to validate implementation of components, thereby ensuring their safe and reliable operation.

The multiagency SBIR support has further allowed the company to expand operations from 5 core employees in 2000 to 15 employees in 2007. Additionally, in 2004,

EDAptive Computing received a $45,000 commercialization assistance award from the NASA Glenn Garrett Morgan Commercialization Initiative to support marketing, planning, and awareness efforts in the defense and aerospace industries.

Product Outcome

EDAstar is ECI's unique solution to completely capture and validate system design requirements. With graphical editing tools, EDAstar-based solutions can be used to rapidly create high-level, high-confidence design concepts with automatically generated tests in a fraction of the time needed by current methods. Further, EDAstar can be used for simulating requirements, assessing risks, and checking their consistency and correctness before expensive mistakes are made in system design and development.

In addition, EDAstar-generated tests, monitoring, and assertions can be used to verify and validate a design or implementation against its specification. EDAstar complements and bridges gaps in existing commercial-off-the-shelf (COTS) tool-based design flows, fitting in the design flow between tools to capture requirements and tools to create detailed specifications and design. Furthermore, EDAstar tools and models can be used as the framework and semantic glue, respectively, for integrating multidisciplinary models, tools, and methods for modeling and simulating a multidisciplinary system of systems.

EDAshield, an additional product offering from ECI, can be used to diagnose, predict, and correct errors after a system has been deployed using EDAstar-created models. EDAshield is a collection of methods and reusable software and hardware assets for system security and can be used to assure trustworthiness, as well as generate anti-tamper logic to protect hardware and software against reverse engineering.

Initial commercialization for the EDAstar product included application by a large prime contractor in a military setting, plus the award of a 5-year U.S. Naval Air Systems Command delivery order contract with a ceiling of over $45 million, entitled "Competent/COTS Upgrade Recertification Environment." Customers include various branches within the U.S. Department of Defense, industry giants like the Lockheed Martin Corporation, Science Applications International Corporation, and Ball Aerospace and Technologies Corporation, as well as NASA's Langley and Glenn Research Centers. ❖

EDAstar™, EDAshield™, and Syscape™ are trademarks of EDAptive Computing Inc.

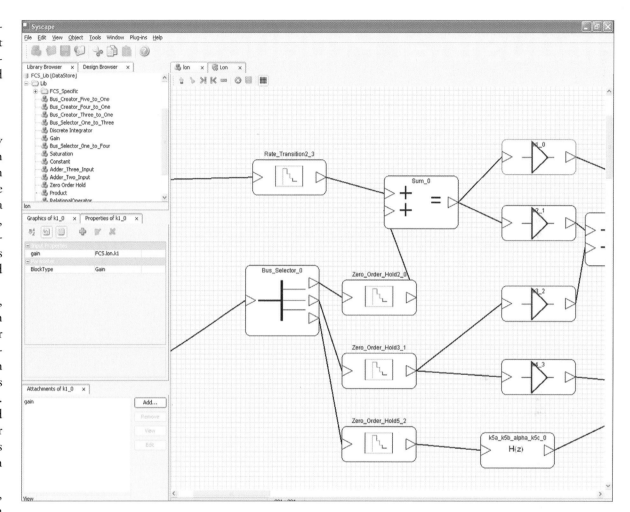

EDAstar can be used for simulating requirements, assessing risks, and checking their consistency and correctness before expensive mistakes are made in the system design and development.

Industrial Productivity

NASA technologies aid industry. The benefits featured in this section:

- Strengthen structures
- Boost data transmission
- Enhance precision fabrication
- Broaden sensing horizons
- Resist extreme heat and stress
- Develop ultra-hard steel
- Save time and energy
- Streamline production
- Control noise and vibration
- Advance thermal management

Open-Lattice Composite Design Strengthens Structures

Originating Technology/NASA Contribution

NASA has invested considerable time and energy working with academia and private industry to develop new composite structures that are capable of standing up to the extreme conditions of space. Over time, such technology has evolved from traditional monocoque designs, in which the skin of a metal structure absorbs the majority of stress the structure is subjected to, to more complex, geometric designs that not only offer strength to counteract stress loads, but also add flexibility.

With an eye to next-generation space-deployable structures, NASA is continuing to identify advanced composite materials and designs that could eventually become the mainstay of future missions, be it in the framework of advanced spacecraft or in the form of extraterrestrial outposts constructed for long-term space habitation.

One of the structures in which NASA has made an investment is the IsoTruss grid structure, an extension of a two-dimensional "isogrid" concept originally developed at McDonnell Douglas Astronautics Company, under contract to NASA's Marshall Space Flight Center in the early 1970s. IsoTruss is a lightweight and efficient alternative to monocoque composite structures, and can be produced in a manner that involves fairly simple techniques.

Partnership

In the early 1990s, NASA's Langley Research Center helped fund a 7-year research project that led to the development of IsoTruss. David Jensen, a professor of civil engineering at Utah's Brigham Young University (BYU) and director of the school's Center for Advanced Structural Composites, invented the technology, with assistance from several graduate students and further funding from the National Science Foundation and the Federal Highway Administration.

Passionate about aerospace, Jensen originally thought that this would be the area in which IsoTruss would attain the most success. The support he received from NASA was for development of space applications, including evaluation of a proposed solar sail that could propel a craft through space by harnessing solar flux. According to him, a very lightweight structure would be required to stretch the solar sail out, because of the high costs associated with launching mass into space.

BYU inked a licensing agreement with space suit and soft materials developer ILC Dover LP in February 2002 to evaluate use of Jensen's patented IsoTruss technology on NASA's conceptual solar sail-driven spacecraft. Though the project never came to fruition, Jensen continues to explore aerospace applications based on his initial support from NASA to create the technology.

Product Outcome

BYU has licensed the IsoTruss technology for use in the United States, China, and Japan; negotiations are in process for other parts of the world.

The technology is garnering global attention, because it is extremely lightweight and as much as 12 times stronger than steel (depending on the application). It is typically constructed of carbon and/or fiberglass filaments that are interwoven in an open-lattice design—longitudinal and helical reinforcement members forming triangles and pyramids that distribute stresses. (The "iso" and "truss" in IsoTruss represent the isosceles triangles that truss the pyramids that ultimately give the structure its strength and stiffness.) According to Jensen, the breakthrough is not the composite materials used to fabricate the structure—as any fiber and resin combination can be used—but the three-dimensional, spider web-like structural design itself, since it eliminates the weight of comparable structures, such as solid wood and tubular metal poles. In essence, a 90-pound IsoTruss composite structure could replace a 1,000-pound steel pole and still offer the equivalent strength.

An IsoTruss structure can be built in many different geometric configurations and possesses many different

The IsoTruss technology delivers superior structural solutions that are lighter, stronger, and more efficient than traditional fiber-reinforced composites, metal, or wood.

geometric variables so that it can be optimally tailored for myriad applications. The open-lattice design enables a variety of standard and innovative connections, while offering significant resistance to column buckling.

There are many cost and environmental advantages to IsoTruss, too: It is less expensive to manufacture, transport, and install than wood or steel; there is little-to-no maintenance needed, as it does not rust, corrode, or rot; it is impervious to insect and woodpecker damage; it is easy to repair and easy to replace, if necessary; it does not contain hazardous or toxic chemicals; and it can be made from fully recyclable materials. When used as utility poles, IsoTruss structures can serve as an environmentally friendly alternative to wooden poles, which are treated with chemical preservatives that could potentially harm the ground and water supplies.

Furthermore, an IsoTruss structure offers superior wind resistance. Because of its open-lattice design, wind drag is typically reduced by 30 percent, as compared to a traditional wooden pole. This means that an IsoTruss pole is much less susceptible to breaking or falling when subjected to high winds—another reason why it is a good match for utility-related applications.

IsoTruss utility pole applications are still on the horizon, albeit with plenty of potential. According to a Wall Street Journal article from 2000, the United States alone requires approximately 4 million new utility poles each year just to replace rotting wood poles, while rapidly developing Third World countries still constructing their power and communications distribution systems will require hundreds of millions of poles over the next decade.

Meanwhile, IsoTruss has taken off in the form of meteorological towers. IsoTruss Structures Inc., a Utah-based licensee of the technology, is selling and installing meteorological towers that are 270-feet high and one-tenth the weight of steel, plus less expensive to manufacture and easier to install than steel towers. These tall towers are designed to handle harsh weather conditions, including radial ice loads up to 1/2-inch thick and strong winds up to 90 miles per hour.

The meteorological towers have been fitted with wind instrumentation and installed near the mouth of Spanish Fork Canyon, Utah; in Arizona and New York; and on U.S. Air Force bases, where they are monitoring meteo-

rological conditions in preparation for the installation of large wind turbines that generate electricity from wind power. Since wind characteristics vary with height, these towers provide more accurate wind speed and meteorological measurements, enabling better prediction of the potential long-term energy efficiency of tall wind turbine towers. At the Spanish Fork site, for example, tests are being conducted to evaluate the economic and energy-saving effects of the tower.

IsoTruss structures can also be employed as preassembled, three-dimensional, corrosion-resistant reinforcement

Civil Engineering students at Brigham Young University built an ultra-strong, super-light mountain bike using carbon fiber IsoTrusses instead of regular cylindrical tubes.

for concrete structures and as standalone structural columns. They were recently used as tilt-up wall braces in constructing the South Towne Exposition Center, in Sandy, Utah. At this site, it took just two workers to install these wall braces (one to hold a brace in position and one to install the anchor), as opposed to the five workers required to install the neighboring steel braces (four to hold and one to install).

IsoTruss is also being considered for use in communication towers, military structures, freeway sign support structures, medical prostheses, farming and irrigation equipment, vehicle and aircraft parts, and sporting goods equipment.

IsoTruss Structures Inc. and ATK Thiokol Propulsion have agreed to collaborate on the analysis and distribution of composite lattice structures for commercial and military products. The pact will enable the two companies to provide application-specific solutions to customers in the military, international aerospace industry, and commercial markets that rely on high-performance, lightweight structures.

In the realm of sports, there is one unique application that may eventually see the light of day as a commercial product. Under the tutelage of Jensen, a group of BYU engineering students used the intertwining IsoTruss composite materials (carbon fiber and Kevlar, in this case) to construct a mountain bike. The longitudinal frame tubes were designed to resist all axial and bending loads, while other IsoTruss parts were used to support transverse sheer and torsion loads. A clear sheathing was placed over some of the tubes to protect them from the dirt and mud that can kick up during a ride. The students involved with the project are touting the bike frame to be lighter, more aerodynamic, and less breakable than many top-of-the-line carbon fiber frames. ❖

IsoTruss® is a registered trademark of Brigham Young University. Kevlar® is a registered trademark of E.I. DuPont de Nemours and Company.

Ultra-Sensitive Photoreceiver Boosts Data Transmission

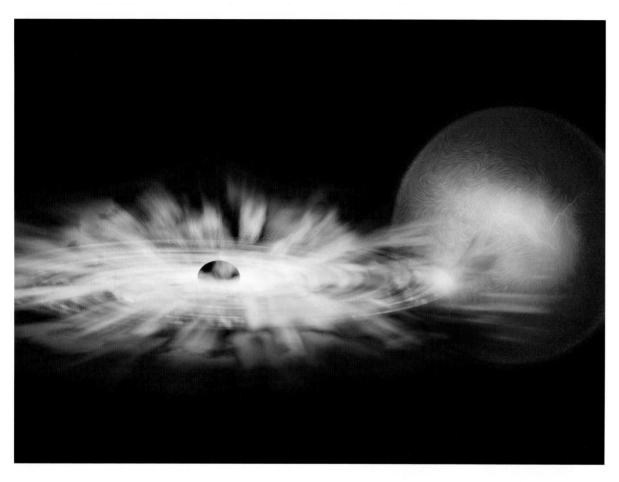

Black holes grow by drawing gas from nearby objects such as stars into an accretion disk, a structure formed when material (usually gas) is being transferred from one celestial object to another.

Originating Technology/NASA Contribution

In June 2006, NASA scientists used extensive data transmitted from the Chandra X-ray Observatory deep space telescope to prove that up to 25 percent of the light illuminating the universe comes from the "massive crush of matter succumbing to the extreme gravity of black holes."

Six months later, in December, the Mars Exploration Rover, Opportunity, photographed patterns in the layering of crater-wall cliffs on Mars that are the clearest evidence of ancient sand dunes the robot has recorded since arriving on the Red Planet in 2004.

In helping to unravel mysteries such as these, NASA depends on advanced, ultra-sensitive photoreceivers and photodetectors to provide high-data rate communications and pinpoint image-detection and -recognition capabilities, respectively, from great distances spanning the universe. These particular parts are now being manufactured with more sensitivity than ever before, so NASA, in staying ahead of the curve, is seeking nothing but the best when it comes to choosing technologies that contribute to the success of its missions.

Partnership

In 2003, Epitaxial Technologies LLC was awarded a **Small Business Innovation Research (SBIR)** contract from Goddard Space Flight Center to help NASA address its need for advanced sensor components and systems for deep space and Mars missions. A Baltimore-based manufacturer and supplier of highly differentiated sensor components for the aerospace, defense, and telecommunications industries, Epitaxial Technologies had previously developed an ultra-sensitive photoreceiver for the U.S. Air Force and the U.S. Department of Defense's Defense Advanced Research Projects Agency (DARPA). According to the company, this photoreceiver was based on the monolithic integration of photodiode detectors, optical amplifiers, and electronic amplifiers on a single chip; possessed a level of sensitivity higher than competing sensors; and could transmit an extremely high rate of data from extremely far distances.

For the NASA SBIR project, Epitaxial Technologies proposed developing an even better-performing version of this photoreciever, which the company already considered to be the best of its kind at the time. With support from Goddard, the company devised a technology that is more sensitive (capable of single photon sensitivity), smaller in size, lighter weight, and requires less power than its predecessor. The resulting technology is intended to boost data rate transmissions well into the terabit range for future space missions. In addition, it has the potential for use in NASA's Earth-based missions for remote sensing of crops and other natural resources.

Product Outcome

Epitaxial Technologies' new monolithic photoreceiver is making a statement on Earth with a wide range of commercial applications. It possesses the ability to operate in several wavelength ranges for fiber optic communications, law enforcement (radar), commercial laser range finding and imaging, and quantum encryption and computing. It also has applications for fluorescence and phosphorescence detection in chemical and biochemical assays.

For fiber optic communications, there is great potential for the ultra-sensitive photoreceiver to boost many military and civilian applications. Although components for high-speed transceivers are widely available, they can be either too expensive, or their bandwidth and distance capabilities can be inadequate, claims Epitaxial Technologies. In sharp contrast, the company says its photoreceiver helps reduce the cost of fiber optic systems and helps enable higher bandwidth and longer distance capabilities suitable for fiber optic networking, especially in metropolitan optical networks and long-haul communications.

There is also great potential for Epitaxial Technologies' photoreceiver in military and civilian active laser imaging and free-space laser communication. These two applications focus on the transmission of laser energy through the atmosphere. During these transmissions, many problems related to signal reception and interference can occur, affecting functions such as laser pointing, tracking, and speckle (light patterns resulting from the reflection of coherent light at rough surfaces), as well as information processing. Typically, electronic amplification of the received optical signal is applied to alleviate these problems; however, this approach can create more noise. Epitaxial Technologies' answer to this is to apply optical amplification techniques at the chip level to increase signal strength with minimum noise amplification and, thus, increase the transmission distance and overcome atmospheric attenuation.

Epitaxial Technologies is also in the business of developing other types of low-cost, ultra-sensitive detection components, so it is always seeking opportunities to further develop high-performance telecommunications and imaging applications. In 2006, the company was again selected to participate in the NASA SBIR program with Goddard. This time, the company is developing monolithic, time-delay photodiode arrays for NASA satellite tracking and vehicle docking, with the intent to spin off another technology with terrestrial benefits. ❖

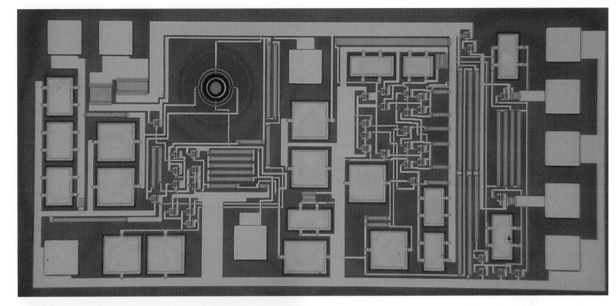

A NASA-funded research project helped Epitaxial Technologies LLC to commercialize its avalanche photodiode and photon counting sensor technologies. Pictured here is a receiver chip.

Top end of a lens for a detector/receiver package.

Micro Machining Enhances Precision Fabrication

Originating Technology/NASA Contribution

In President Ronald Reagan's 1984 State of the Union address, he announced plans for a U.S. space station, the equivalent of the Russian space station, Mir. This announcement set off a flurry of congressional funding debates, and it was not until 1988 that the President announced that a consensus had been reached and the project would go forward. The project was named "Space Station Freedom."

It was to be a permanently manned, Earth-orbiting outpost that could serve as a repair shop for the shuttle fleet, a microgravity laboratory, an observation point for astronomers, and an assembly station for spacecraft heading even farther into space. The project came to an end, unrealized, in 1990, when it was revealed that the design was over budget, overweight, and had been so complicated and compromised by the political debates and budget restrictions that it could no longer be realized.

Although Freedom never made it past the design stages, the science, and many aspects of the original designs, made their way onto the International Space Station (ISS), whose first component was launched in 1998. For instance, several advanced thermal systems developed for cooling everything from battery components to crew cabins originally designed for Space Station Freedom are still in use on the ISS.

These thermal systems, while more advanced and specialized than most used on Earth, like many terrestrial refrigerators, employ evaporative ammonia as their coolant. To create this simple refrigerant, heat is applied to a mixture of water and ammonia until the solution reaches the boiling point of the ammonia. The boiling solution then flows to another chamber, where the ammonia gas separates from the water. The gas floats upwards to a condenser, where a series of fins and coils cool it, allowing it to condense back into a liquid. The liquid ammonia then flows to an evaporator, where it is mixed with hydrogen

Artist's concept of the Space Station Freedom, a design that led to many of the technologies currently aboard the International Space Station.

gas, and, when it evaporates, produces cold temperatures. Having fulfilled its refrigerating purposes, the ammonia, along with the hydrogen, mixes again with the water. This solution releases the hydrogen gas, where it returns to await the ammonia gas, as the cycle continues.

Even though this same series of chemical reactions is used in space-bound refrigerators as in terrestrial versions, the space-bound coolers must have one major difference: they must be significantly smaller. Refrigerators are notoriously heavy, and weight is always a payload concern for space-bound equipment, so the Space Agency needed to engineer smarter, more efficient thermal systems.

Partnership

In the 1980s, through two **Small Business Innovation Research (SBIR)** contracts with Johnson Space Center, Dr. Javier Valenzuela worked on a project to develop an ammonia evaporator for thermal management systems aboard Freedom. At the time, he was serving as the principal investigator for these contracts while working with Creare Inc., of Hanover, New Hampshire.

In 1991, Valenzuela formed Mikros Technologies Inc., based in Claremont, New Hampshire, to commercialize the work he had done under the NASA contracts. In

2001, the company was awarded two SBIR research contracts from Goddard Space Flight Center and is, to this day, actively engaged in advancing micro-fabrication and high-performance thermal management technologies.

Product Outcome

The technique Valenzuela developed was an advanced form of micro-electrical discharge machining (micro-EDM) to make tiny holes in the ammonia evaporator. The evaporator relied on droplet impingement cooling to achieve high heat flux and low thermal resistance. Many thousands of small nozzles were required to "print" a thin layer of ammonia over the surface of the evaporator. However, no technology existed at that time for the fabrication of suitably shaped micro-nozzles in metals. Mikros' micro-EDM technology was developed to meet this need. Micro-EDM is an erosion process, in which electrical discharges between two electrodes (one on the tool, and the other on the conductive work surface) machine very small holes, channels, and cavities.

Micro-EDM can cut complex shapes into a variety of conductive materials, regardless of the material's hardness, even hard materials such as steels and carbides. It can also be used on materials like ferrites and silicon, which have the tendency to crack or become brittle when exposed to traditional (macro) EDM.

All EDM has the additional advantage of noncontact machining. The method, which relies on a gap between the tool and work surface for the discharging of the spark, allows the process to take place with no pressure being placed on the material. This means that micro-EDM can be used on very thin and very fragile surfaces, and even on curved surfaces, without damaging them.

Mikros has had great success applying this method to the fabrication of micro-nozzle array systems for industrial ink jet printing systems. The company is currently the world leader in fabrication of stainless steel micro-nozzles for this market. It routinely fabricates nozzle arrays with hundreds or even thousands of shaped nozzles with diam-

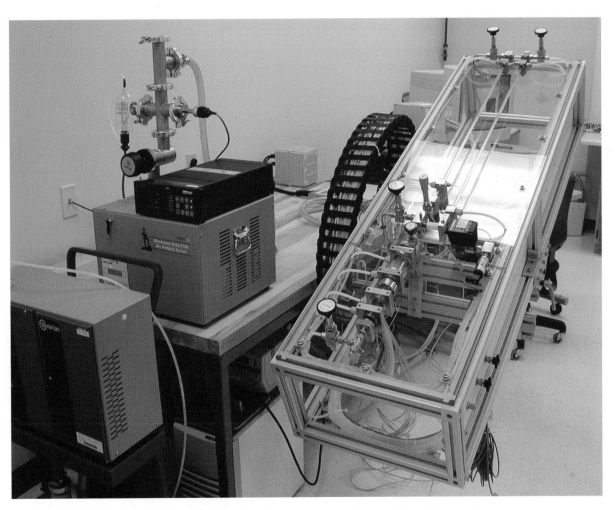

Breadboard and test station of Mikros' High-Heat Flux Evaporator for the Phase II NASA SBIR project.

eters ranging from 30 to 80 microns. The micro-nozzle arrays are machined in thin, free-standing foils, or on foils that are diffusion bonded to thicker substrates with complex internal flow passages.

The company has continued its relationship with the Space Agency, too. Its other division, in addition to micro-EDM, is thermal systems, which is currently still involved with Goddard through the SBIR contracts to develop a high-heat flux capillary evaporator for use in loop heat pipes or capillary pumped loops. ❖

Portable Hyperspectral Imaging Broadens Sensing Horizons

Originating Technology/NASA Contribution

All objects reflect a certain amount of energy, even if it is just the electromagnetic energy created by the movement of electrically charged molecules. Measurements of these reflected energies, called spectra, can be used to create images of observed items and can thus serve to identify objects and substances. To create a spectral image, the intensity of the energy an object is reflecting is measured at different wavelengths, and then these measurements are assigned colors.

These images can be simple, as they often are with broadband or multispectral imaging, in which an image may consist of very few measurements. An example of this would be the type of pictures created by infrared or "night vision" equipment. This broadband multispectral imaging can be very helpful in showing differences in energy being radiated and is often employed by NASA satellites to monitor temperature and climate changes.

Images that are created with hundreds to thousands of spectra, hyperspectral images, allow for more subtle features to be shown, including the distinction between natural and man-made objects. Hyperspectral imaging is ideal for advanced laboratory uses, biomedical imaging, forensics, counter-terrorism, skin health, food safety, and Earth imaging.

Engineers at NASA's Stennis Space Center work with the Institute for Technology Development (ITD), a nonprofit research company located onsite at Stennis, to develop new hyperspectral imaging technologies. ITD has been at the forefront of imaging technologies since it opened nearly 20 years ago, revolutionizing the fields of high-resolution imaging and measurement technology, with a focus on transferring this newfound technology to industry. Recent work led to reduced sensor sizes, elimination of the need for either the sensor or target to be in motion in order to obtain images, and the development of a portable hyperspectral imaging device applicable for a wide variety of uses.

Partnership

The portable device created by the NASA-ITD partnership was licensed, and in 2003, a new spinoff company, Photon Industries Inc., was formed. In 2005, ITD and its start-up brought the technology and business plan to the World Technologies Venture Capital Exposition, held that year in Texas, where it was discovered by Lextel Intelligence Systems LLC, of Jackson, Mississippi. In November 2005, Lextel purchased the company and its cutting-edge NASA technology.

The acquisition was a great boon for the technology, as it now had access to worldwide marketing, and later that year, it was inducted into the Space Technology Hall of Fame, created by the Space Foundation, in cooperation with NASA, to increase public awareness of the benefits that result from space exploration programs and to encourage further innovation.

Product Outcome

Since the acquisition, in addition to winning awards, Lextel has also added new features to and expanded the applicability of the hyperspectral imaging systems. It has made advances in the size, usability, and cost of the instruments. The company now offers a suite of turnkey hyperspectral imaging systems based on the

Lextel Intelligence Systems LLC made advances in the size, usability, and cost of the NASA-developed portable hyperspectral imaging instruments. The company now offers a suite of turnkey hyperspectral imaging systems based on the original NASA groundwork.

original NASA groundwork. It currently has four lines of hyperspectral imaging products: the EagleEye VNIR 100E, the EagleEye SWIR 100E, the EagleEye SWIR 200E, and the EagleEye UV 100E.

The EagleEye VNIR 100E spans the spectrum from the visible to infrared, measuring the reflected energy of the target at hundreds of narrow wavelengths. It incorporates a patented line-scanning technique that eliminates the need for relative movement between the target and the sensor, which eliminates the need for liquid crystal tunable filters, which suffer from relatively low throughput. The EagleEye uses a prism-grating prism to separate incoming light into component wavelengths, a method that loses less energy and is, therefore, especially useful in situations where high signal-to-noise ratios are important. Each image created by the VNIR 100E contains a complete reflectance spectrum from 400 to 1,000 nanometers for every picture element in the image.

The next product in Lextel's suite of hyperspectral imaging instruments is the EagleEye SWIR 100E.

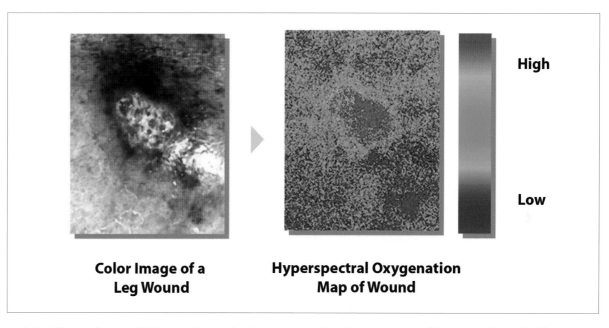

Color Image of a Leg Wound **Hyperspectral Oxygenation Map of Wound**

Lextel Intelligence Systems LLC has distributors for its hyperspectral imaging instruments all throughout the world. They are currently used for a wide variety of applications including medical, military, forensics, and food safety uses.

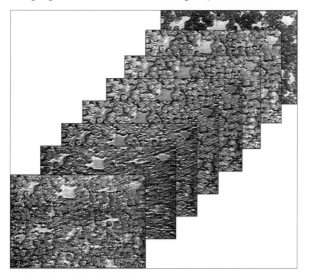

Microscope slide examined using a hyperspectral imaging instrument.

Each image created by this device contains a complete reflectance spectrum from 900 to 1,700 nanometers for every picture element in the image.

The third product in the suite is the EagleEye SWIR 200E. Each image created by this device contains a complete reflectance spectrum from 1,000 to 2,500 nanometers for every picture element in the image.

Lextel's fourth product is the EagleEye UV 100E, an instrument that will span the spectrum from ultraviolet to thermal infrared. With this instrument, the images contain a complete reflectance spectrum from 200 to 400 nanometers for every picture element in the image.

Lextel has distributors for these hyperspectral imaging instruments in the United States, Canada, Europe, China, Japan, and Israel. They are currently used worldwide for a wide variety of applications including medical, military, forensics, and food safety, to name just a few.

In March 2006, Lextel entered into an agreement with ITD and NASA to continue to improve upon the already advanced sensing systems. The three entities are currently working on methods for incorporating hyperspectral fluorescence into the instruments.

In November 2006, Lextel also entered into an agreement with Headwall Photonics Inc., of Fitchburg, Massachusetts, to build a product line based on the convex grating reflective spectrograph design. This will allow the worldwide customer base to have a choice between prism-grating prism systems or reflective-based systems. There will also be work between the two groups to incorporate their microspectrographs into Lextel's next generation system, The Micro Intelligent Imaging System. ❖

Hypersonic Composites Resist Extreme Heat and Stress

Originating Technology/NASA Contribution

On October 14, 1947, Captain Charles "Chuck" Yeager made history when he became the first pilot in an officially documented flight to ever break the sound barrier. Flying a Bell XS-1 test jet over the Mohave Desert, Yeager hit approximately 700 miles per hour, when a loud boom thundered across the barren landscape as he crossed from subsonic to supersonic speeds. The sonic boom, akin to the wake of the plane's shockwaves in the air, occurred at Mach 1—the speed of sound (named after Ernst Mach, an Austrian physicist whose work focused on the Doppler effect and acoustics).

This eventually led to the famous 3-hour trans-Atlantic flights of the *Concorde*, traveling at Mach 2, and the development of fighter jets, which began routinely crossing the sound barrier—the era of supersonic flight. At NASA, however, aerospace engineers wanted to go faster—hypersonic or Mach 5; five times the speed of sound. They continued to push the limits of speed, setting a world record in October 1967, which held for 35 years at Mach 6.7, in the NASA-designed X-15 aircraft.

Never content to rest on its laurels, the Space Agency wanted to go even faster. NASA began the X-43A project,

The X-43A flights were the first actual flight tests of an aircraft powered by a scramjet engine capable of operating at hypersonic speeds (above Mach 5, or five times the speed of sound).

the purpose of which was to go Mach 10. The "X" in the name, as in X-1 and X-15, signifies that these are experimental aircraft, not intended for mass production, but built solely for flight research, and the 43 indicates that it was the 43rd such aircraft.

The major experimental feature of this particular aircraft was a scramjet engine, which kicks in at about Mach 6. In a normal jet engine, blades compress the air, but in the scramjet, the combustion of hydrogen fuel in a stream of air is compressed by the high speed of the aircraft itself. To achieve Mach 10, the X-43A was first carried up into the air attached to a B-52 and then "jumpstarted" with a Pegasus rocket.

It was a multiyear experiment and involved building three test vehicles. The first two were meant to reach speeds of up to Mach 7, and the third was reaching for double digits: Mach 10. NASA hoped that these three unmanned crafts would allow aerospace engineers to further advance understanding of hypersonic flight and that the lessons learned could be applied to increase payload capacity for future vehicles, including hypersonic aircraft and reusable space launchers.

Tullahoma, Tennessee-based ATK-GASL built the X-43A and its engine, while the Huntington Beach, California-based Boeing Company's Phantom Works designed the thermal protection and onboard system. The booster, a modified Pegasus rocket, was built by Orbital Sciences Corporation at the company's facility in Chandler, Arizona. The program was jointly managed by NASA's Dryden Flight Research Center and Langley Research Center.

The first X-43A dropped from the B-52 carrier aircraft on June 2, 2001 and spun out of control when the booster rocket failed to operate properly. It was destroyed

The first X-43A hypersonic research aircraft and its modified Pegasus booster rocket were carried aloft by NASA's B-52 carrier aircraft from Dryden Flight Research Center on June 2, 2001, for the first of three high-speed free flight attempts.

and allowed to fall into the ocean. The second test vehicle launch was a success, making the X-43A the fastest free-flying, air-breathing aircraft in the world. On November 16, 2004, NASA launched the third X-43A scramjet aircraft, which reached record-breaking speeds near Mach 9.8 at an altitude of around 110,000 feet. The aircraft still holds the Guinness World Record.

Partnership

Through research contracts with NASA, Materials and Electrochemical Research Corporation (MER), of Tucson, Arizona, contributed a number of technologies to these record-breaking flights. Over the course of 6 years, MER completed Phase II **Small Business Innovation Research (SBIR)** contracts with Goddard Space Flight Center, Langley Research Center, the Jet Propulsion Laboratory, Johnson Space Center, and Marshall Space Flight Center. By partnering with NASA through these contracts, MER developed a coating that successfully

render the high-thermal conductivity. The parts were first processed in the form of a C-C billet, and then machined to NASA specifications. Oxidation protection was achieved by a dual chemical vapor reaction and chemical vapor deposition process.

Product Outcome

To take advantage of commercializing these specialized composites, MER patented its C-C composite process and then formed a spinoff company, Frontier Materials Corporation (FMC), in Tucson, Arizona. FMC is using the patent in conjunction with low-cost PAN (polyacrylonitrile)-based fibers to introduce these materials to the commercial markets.

The C-C composites are very lightweight, yet still have great strength and stiffness, even at very high temperatures. They can be produced with either low- or high-thermal conductivity, and when graphitized, they have superior electrical conductivity. Even with all of these characteristics, the carbon composites are still relatively inexpensive.

The composites have been used in industrial heating applications, the automotive and aerospace industries, as well as in glass manufacturing and on semiconductors. C-C composites have been used in industrial heating fixtures in the form of structural flat panels, structural members for panel support, and for nuts and bolts. In the automotive industry, the composites have been useful for engine components, brakes, cylinder liners, and panels. Aerospace applications have included missile bodies, leading edges, and structured components. Applications also include transfer components for glass manufacturing and structural members for carrier support in semiconductor processing. ❖

MER's carbon-carbon composites are very lightweight, yet still have great strength and stiffness, even at very high temperatures.

passed testing at simulated Mach 10 conditions, and provided several carbon-carbon (C-C) composite components for the flights.

MER created all of the leading edges for the X-43A test vehicles at Dryden. Considered the most critical parts of this experimental craft, the leading edges had several specific requirements. As the vehicle's speed increased, so did heat and thermal load, approaching 4,000 °F, well above the temperatures the shuttle is exposed to during reentry. In addition to being very heat resistant, the coating had to be very lightweight and thin, as the aircraft was designed to very precise specifications and could not afford to have a bulky coating.

In total, 11 C-C leading edges were created for the nose, the chines (the areas where the body meets the wing), the verticals, and 2 horizontals. These parts were made with P-30X graphite fibers, using a liquid matrix process. A very high process temperature was utilized to

Computational Modeling Develops Ultra-Hard Steel

Originating Technology/NASA Contribution

Glenn Research Center's Mechanical Components Branch routinely conducts research on transmissions and gearing for advanced gas turbines, promoting their safety, weight reduction, and reliability. The Mechanical Components Branch is staffed by both NASA and U.S. Army Research Laboratory employees, and the research program is designed and executed to meet the needs of both organizations. The researchers have developed a world-class set of instruments and test devices, including a spiral bevel or face gear test rig for testing thermal behavior, surface fatigue, strain, vibration, and noise; a full-scale, 500-horsepower helicopter main-rotor transmission testing stand; a gear rig that allows fundamental studies of the dynamic behavior of gear systems and gear noise; and a high-speed helical gear test for analyzing thermal behavior for rotorcraft. These are just a few examples of the highly specialized equipment the researchers at the Mechanical Components Branch have at their disposal.

Since 1972, the branch's spur gear fatigue rig has set the standard for gear surface fatigue experiments, enabling development of robust, efficient, and safe gas turbines and rotorcraft. The test rig provides accelerated fatigue life testing for standard spur gears at speeds of up to 10,000 rotations per minute, and enables engineers to investigate the effects of materials, heat treat, shot peen, lubricants, as well as other factors, on the gear's performance.

Partnership

QuesTek Innovations LLC, based in Evanston, Illinois, is an innovative materials solutions company that designs and develops new materials in less than 50 percent of the time and at less than 30 percent of the cost of traditional empirical methods.

QuesTek provides unique materials solutions to a variety of customers by using its powerful mechanistic

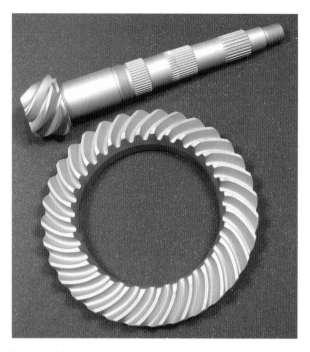

Uses for this new class of steel are limitless in areas needing exceptional strength for high-throughput applications.

computational models to design steels and alloys to customer requirements. The company's methodology uniquely allows for design trade-offs, such as steels optimally balanced for strength, toughness, and corrosion resistance. Employing computational materials design techniques, QuesTek uses fundamental thermodynamic and kinetic data to predict alloy microstructures; in contrast to traditional empirical—trial and error—approaches.

QuesTek recently developed a carburized, martensitic gear steel with an ultra-hard case using its computational design methodology, but needed to verify surface fatigue, lifecycle performance, and overall reliability.

The Battelle Memorial Institute, a nonprofit global science and technology enterprise that develops and commercializes technology and manages laboratories for customers, introduced the company to researchers at Glenn's Mechanical Components Branch and facilitated a partnership allowing researchers at the NASA Center to conduct spur gear fatigue testing for the company.

Testing revealed that QuesTek's gear steel outperforms the current state-of-the-art alloys used for aviation gears in contact fatigue by almost 300 percent. This testing generated necessary data to demonstrate the superiority of the ultra-hard case alloy and quantified the contact fatigue benefit attainable using QuesTek's material. However, bending fatigue testing demonstrated that the bending fatigue capability of the new steel was somewhat less than the capability of the current state-of-the-art alloys used for aviation gears. The test data has provided guidance for engineers to select the best combination of properties to satisfy the requirements of a given application.

Product Outcome

With the confidence and credibility provided by the NASA testing, QuesTek is commercializing two new steel alloys. These alloys combine maximum case hardness with a tough, ductile core, promoting high wear and contact fatigue life and offer a 20-percent increase (or more) in gear endurance in high-power density aerospace transmission systems.

Uses for this new class of steel are limitless in areas needing exceptional strength for high-throughput applications. The material is already being used in racing markets. For instance, QuesTek's C61 material was the power behind the Class 1600 champion in last year's Baja 1000 off-road race. Aside from racing, this high-performance material is in testing for heavy equipment gearing and in oil/gas gearing applications. NASA is interested in the materials for use in developing vertical takeoff and landing vehicles, based on conventional rotorcraft but with the speed and high-altitude performance of turbo propellers. Advancements in gear design and reliability make possible the development of these highly specialized vehicles. ❖

Thin, Light, Flexible Heaters Save Time and Energy

Originating Technology/NASA Contribution

Ice accumulation is a serious safety hazard for aircraft. The presence of ice on airplane surfaces prevents the even flow of air, which increases drag and reduces lift. Ice on wings is especially dangerous during takeoff, when a sheet of ice the thickness of a compact disc can reduce lift by 25 percent or more. Ice accumulated on the tail of an aircraft (a spot often out of the pilot's sight) can throw off a plane's balance and force the craft to pitch downward, a phenomenon known as a tail stall.

The Icing Branch at NASA's Glenn Research Center uses the Center's Icing Research Tunnel (IRT) and Icing Research Aircraft, a DeHavilland Twin Otter twin-engine turboprop aircraft, to research methods for evaluating and simulating the growth of ice on aircraft, the effects that ice may have on aircraft in flight, and the development and effectiveness of various ice protection and detection systems.

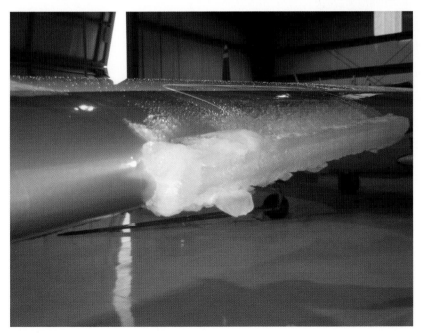

Developed in conjunction with NASA for in-flight aircraft deicing, Q•Foil heaters are now available for a broad range of applications from as small as a single square inch to as large as several square feet, and provide extremely rapid thermal response and even heating for a wide array of temperature ranges.

Partnership

EGC Enterprises Inc. (EGC), of Chardon, Ohio, used the IRT to develop thermoelectric thin-film heater technology to address in-flight icing on aircraft wings. Working with researchers at Glenn and the original equipment manufacturers of aircraft parts, the company tested various thin, flexible, durable, lightweight, and efficient heaters and developed a thin-film heater technology that they discovered can be used in many applications, in addition to being an effective deicer for aircraft.

Product Outcome

The result of this research was the development of a new thermoelectric heater the company has dubbed the Q•Foil Rapid Response Thin-Film Heater, or Q•Foil, for short. The product meets all criteria for in-flight use and promises great advances in thin-film, rapid response heater technology for a broad range of industrial applications. Primary advantages include time savings, increased efficiency, and improved temperature uniformity.

EGC makes the heaters out of thin layers of varying materials that are bonded to form heater laminates. The inner layer, made of a flexible graphite foil (marketed separately by EGC as ThermaFoil), is electro-thermally conductive and typically laminated between a heat-conducting outer layer, and a protective insulating layer. The inner layer provides full electrical conductivity, eliminating the need for wire elements, metal etchings or heat-conductive fibers. In addition to being readily available and relatively inexpensive, the graphite foil is energy efficient.

Because Q•Foil conducts heat well and can cover a large area, little energy is needed for it to raise the surface temperature to the necessary degree, and it is capable of heat increases as rapid as 100 °F per second if needed, which can translate into savings in time or energy. Designers can configure it to heat an item more quickly or to heat an item to the same temperature as other heaters would, while expending less energy.

Not only does Q•Foil work quickly and efficiently, it is also precise, controllable to within 3 °F. If it is being used for an application that requires different temperature zones, Q•Foil can also be configured to accommodate this within the same heating coil.

While the product is available in a variety of sizes, ranging from 1 square inch to 100-foot lengths, the company notes that it is most efficient over large spaces, as this allows the user to get the most advantage of its unique cost- and energy-saving properties. Thin-film flexibility allows it to be mounted to a variety of objects, and it maintains flexibility through a full range of temperatures. The company has services to assist customers with determining the right dimensions and designs they will require for a specific job.

In addition to wing deicing, EGC has begun looking at the material's usefulness for applications including cooking griddles, small cabinet heaters, and several laboratory uses. ❖

Q•Foil™ is a trademark, and ThermaFoil® is a registered trademark of EGC Enterprises Inc.

Novel Nanotube Manufacturing Streamlines Production

Originating Technology/NASA Contribution

Nanotubes are sheets of graphite, one atom thick, rolled into seamless cylinders, with an exterior diameter in the range of nanometers. For a sense of perspective, if you were to split a human hair into 50,000 independent strands, a nanotube would be about the size of one of those strands.

What would someone do with anything that small? These nanostructures have novel qualities that make them uniquely qualified for a plethora of uses, including applications in electronics, optics, and other scientific and industrial fields.

These thin, hollow tubes have remarkable strength, especially considering their microscopic size. Stronger than steel, nanotube strands can be used to form extremely strong, yet lightweight, materials. They are efficient conductors of heat (able to withstand temperatures up to 2,000 °C in the absence of oxygen) and also possess unique electrical properties. Nanotubes can be manufactured so that they conduct electricity as well as copper, but can also be made to function as semiconductors (able to switch from conducting electricity to insulating from it), making them quite valuable in the production of miniaturized electronic components.

Although scientists have been aware of these nanostructures since their discovery in 1991 by Japanese physicist Dr. Sumio Iijima, practical use has been thwarted by the high costs, complexity, and even danger of manufacturing them. A group of researchers at Goddard Space Flight Center, led by Dr. Jeannette B. Benavides, however, developed a manufacturing process for single-walled carbon nanotubes (SWCNTs) that overcomes these obstacles.

Typical methods for creating these nanostructures, whether chemical vapor deposition, laser ablation, microwave, or high-pressure carbon monoxide conversion, use metal catalysts to encourage the carbon to grow into the tube shape, as opposed to capping and sealing. The NASA process uses helium arc welding to vaporize an amorphous carbon rod and then form nanotubes by depositing the vapor onto a water-cooled carbon cathode, which then yields bundles, or ropes, of single-walled nanotubes at a rate of 2 grams per hour using a single setup.

The NASA-developed process eliminates the costs associated with the use of metal catalysts, including the cost of product purification, resulting in a relatively inexpensive, high-quality, very pure end product. The process employs an arc welder, a helium purge, an ice water bath, and basic processing experience. This significantly simplifies previous complex, dangerous manufacturing processes, which required expensive equipment like vacuum chambers, dangerous gasses, and extensive technical knowledge.

While managing to be less expensive, safer, and simpler, the process also increases the quality of the nanotubes. Since no metal catalyst is used, no metal particles need to be removed from the product. The elimination of these metal impurities increases the temperature at which the materials will degrade and eliminates any damage that may have been caused by purification processes used to remove metal residue.

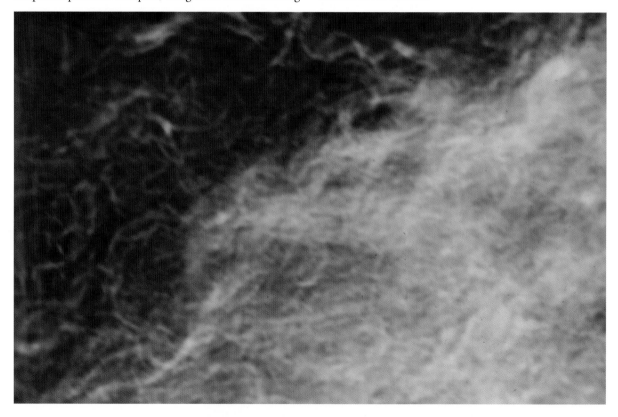

NASA's single-walled carbon nanotube (SWCNT) manufacturing process eliminates the costs associated with the use of metal catalysts, including the cost of product purification. As a result, the manufacturing cost can be reduced significantly for high-quality, very pure SWCNTs.

Partnership

In 2003, Benavides filed a New Technology Report at Goddard, bringing her invention to the attention of the Center's Innovative Partnerships Program (IPP) office, which acknowledged the role this production process could play in making carbon nanotubes more accessible. The IPP office started the process of finding commercial partners.

To get this cutting-edge process into the hands of the public, Goddard's IPP office promoted the technology in print, online, and at industry conferences. In 2005, the technology captured the interest of Wayne Whitt, an entrepreneur interested in forming an advanced materials corporation, but who was in search of an innovation that would help his company stand out from the crowd. Shortly after learning about the SWCNT manufacturing technology, Whitt applied for a nonexclusive license and formed the Boise-based Idaho Space Materials Inc. (ISM).

Product Outcome

Once the license was granted, ISM then worked with Benavides and the University of Idaho's Electron Microscopy Center to examine and improve upon variables in the manufacturing process, ultimately enhancing the process by increasing yield and production capacity. Once the process was tweaked, ISM was ready to commercialize its products, and the inexpensive, robust nanotubes are now in the hands of the scientists who will create the next generation of composite polymers, metals, and ceramics that will impact the way we live. In fact, researchers are examining ways for these newfound materials to be used in the manufacture of transistors and fuel cells, large screen televisions, ultra-sensitive sensors, high-resolution atomic force microscopy probes, super-capacitors, transparent conducting films, drug carriers, catalysts, and advanced composite materials, to name just a few.

Unlike most current methods—which require expensive equipment (e.g., vacuum chamber), dangerous gasses, and extensive technical knowledge to operate—NASA's simple SWCNT manufacturing process needs only an arc welder, a helium purge, an ice water bath, and basic processing experience.

In August 2006, ISM unveiled its new line of single-walled nanotubes with no metal catalyst under the name NOMEC 1556. The advanced materials company has had tremendous success with its new product, and has nearly tripled in size.

The transfer of this NASA technology has also been beneficial to NASA, as widespread use of the Space Agency's patented technology will produce revenue that NASA can reinvest in additional research, and it now has a reliable source from which to obtain high-quality, low-cost SWCNTs for use in its research and missions. ❖

'NASA Invention of the Year' Controls Noise and Vibration

Originating Technology/NASA Contribution

Developed at NASA's Langley Research Center, the Macro-Fiber Composite (MFC) is an innovative, low-cost piezoelectric device designed for controlling vibration, noise, and deflections in composite structural beams and panels. It was created for use on helicopter blades and airplane wings as well as for the shaping of aerospace structures at NASA.

The MFC is an actuator in the form of a thin patch, almost like a 3- by 2-inch bandage comprised of piezoelectric fibers, an epoxy matrix, and polyimide electrodes, and is also called a piezocomposite. If one applies a voltage to the MFC it will stretch, and if attached to a structure it will cause the surface to bend. The major advantages of a piezofiber composite actuator are higher performance, flexibility, and durability, compared to a traditional piezoceramic actuator.

MFCs consist of rectangular piezoceramic rods sandwiched between layers of adhesive film containing tiny electrodes that transfer a voltage directly to and from ribbon-shaped rods that are no thicker than a few tenths of a millimeter. These miniscule actuators are roughly equivalent to human muscles—flexing, stretching, and returning to their original position when electricity is applied.

Because any external mechanical deformation of an MFC package produces a charge on the electrodes proportional to the deflection, its compression or stretching also enables the MFC to be used as a self-powered sensor. Defects within structures can therefore be detected, or small amounts of energy can be collected and stored for later use.

The MFC's combination of small size, durability, flexibility, and versatility allows it to be integrated—along with highly efficient electronic control systems—into a wide range of products. Potential applications include sonar; range-measuring and fish-finding equipment; directional-force and fingerprint sensors; flow meters; and

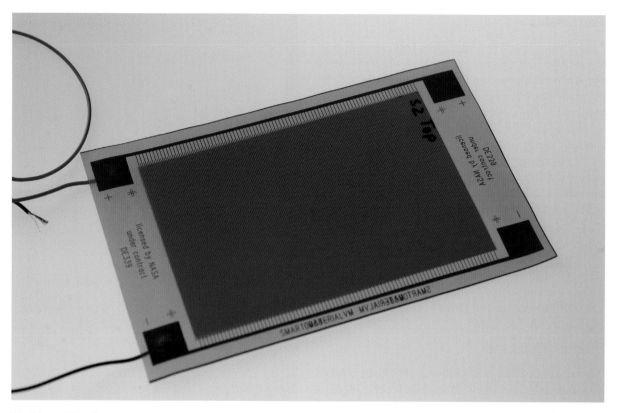

The Macro-Fiber Composite's flat profile and use as a sensor and an actuator allows for use in critical or tight areas where other technologies with larger volumetric profiles cannot be used.

vibration/noise control in aircraft and automobiles. Since its original development in 1999, the MFC has been used in government and industrial applications ranging from vibration reduction to structural health monitoring.

NASA has used MFC piezocomposites for alleviating tail buffeting in aircraft, controlling unsteady aerodynamics and noise on helicopter rotor blades, and actively reducing vibrations in large deployable spacecraft structures. The MFC has been used as a sensor for impedance-based health monitoring of launch tower structures at NASA's Kennedy Space Center, for strain feedback sensing and control in industrial arc welding

equipment, in an STS-123 experiment, in solar sail technology, and in ultra-lightweight inflatable structures.

The MFC has been internationally recognized for its innovative design, receiving two prestigious "R&D 100" awards in 2000, including the "R&D Editor's Choice" award as one of the 100 most significant technical products of the year. The MFC was also the recipient of the International Forum's prestigiowus "iF Gold" award, in Germany, for design excellence in 2004. In March 2007, the MFC was awarded the title of "NASA Invention of the Year."

Partnership

Smart Material Corporation, of Sarasota, Florida, specializes in the development of piezocomposite components. The company licensed the MFC technology from Langley in 2002, and then added it to their line of commercially produced actuators.

It now combines the Langley MFC's piezoelectric properties with the robustness and conformability of plastics to radically extend the spectrum of commercial applications.

A NASA partnership gives a small company access to research and technologies that allow it to compete with larger corporations. According to Thomas Daue, with Smart Material Corporation, "For a small business, it is almost not possible anymore to spend the money for basic research in high tech products. Licensing technology from the leading research facilities in this country is a very cost effective way to become a player in a new technology field. Many developments ready for licensing at government-owned facilities have already reached the proof of concept status, which would often cost a small business or start-up millions of dollars."

Smart Material Corporation is now marketing MFCs internationally, with the majority of applications in the United States directed at Federal government research projects or defense-related government contracts. For example, Smart Material Corporation currently sells the materials to Langley, NASA's Jet Propulsion Laboratory, and Marshall Space Flight Center, where they are used as strain gauge sensors, as well as to the U.S. Air Force and the U.S. Army.

Product Outcome

To date, Smart Material Corporation has sold MFCs to over 120 customers, including such industry giants as Volkswagen, Toyota, Honda, BMW, General Electric, and the tennis company, HEAD. The company also esti-

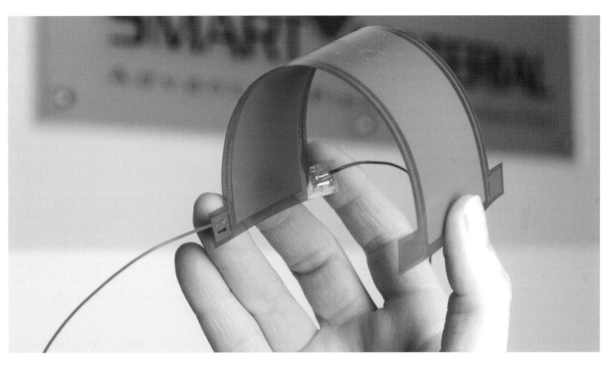

When compared to standard piezoelectric systems, the MFC is much more durable and provides increased unidirectional control. Furthermore, the MFC is designed to be readily integrated into a system as an add-on component or integrated during manufacture.

mates that its customers have filed at least 100 patents for their unique uses of the technology.

Smart Material Corporation's main manufacturing facilities are located in Dresden, Germany, in the vicinity of the Fraunhofer Institute for Ceramic Technologies and Sintered Materials, one of the world's leading research institutes in the field of advanced ceramics. Dresden is also the center of the German semiconductor industry, and so provides crucial interdisciplinary resources for further MFC refinement. In addition to its Sarasota facility, the company also has sales offices in Dresden and in Tokyo, Japan.

The company's product portfolio has grown to include piezoceramic fibers and fiber composites, piezoceramic actuators and sensors, and test equipment for these products. It also offers a compact, lightweight power system for MFC testing and validation.

Smart Material Corporation believes that solid-state actuator systems, including piezoceramics, will have healthy commercial growth in the coming years, with increasing penetration in industrial, medical, automotive, defense, and consumer markets. Consumer applications already on the market include piezoelectric systems as part of audio speakers, phonograph cartridges and microphones, and recreational products requiring vibration control, such as skis, snowboards, baseball bats, hockey sticks, and tennis racquets. ❖

Thermoelectric Devices Advance Thermal Management

Originating Technology/NASA Contribution

When NASA programs need the ultimate reliability to power deep space probes, they repeatedly select thermoelectric (TE) devices as a system component. TE devices heat, cool, and generate electricity when a temperature differential is provided between the two module faces. Using radioactive isotope Plutonium 238 and TE devices to convert waste heat into electricity, NASA has depended on radioisotope thermoelectric generators (RTGs) in 25 U.S. missions since 1961.

NASA relied on RTGs launched in Apollo missions to the Moon, the Viking missions to Mars, and the Pioneer, Voyager, Ulysses, Galileo, and Cassini missions to the outer solar system. More than 30 years from its launch, and well beyond the orbit of Pluto, the RTGs on the Pioneer 10 spacecraft continue to operate.

A project like the Pioneer 10 RTG is an example of the rigorous standards imposed by NASA product specifications and long-life applications that have been known to validate the unique characteristics of TE devices. Like the RTG as a system, the TE device as a component exhibits resilient characteristics. Not only do they have proven long-life performance reliability, they also operate in a vacuum, withstand rigorous vibration, and are relatively insensitive to radiation and other environmental factors.

With confidence, TE devices were integrated into a myriad of niche applications by commercial enterprise. In the 1960s, there was a small handful of TE device manufacturers; in response to wide product utilization, today TE devices are readily available, manufactured by companies around the world.

With the legacy of reliability documented by successful NASA project use, there has been a proliferation of commercialized products using TE devices, and these components have been used in pacemakers, undersea defense and communication systems, and in Arctic weather stations. Where conventional compressor and chemical cooling methods are impractical, TE devices can be switched from coolers to heaters and operate over a broad range of power, while also able to be more temperature precise within tight tolerances than traditional systems. These qualities make TE devices perfect for integration in a wide array of temperature control applications like miniature infrared detectors, circuits in cruise missiles and aircraft, lasers, blood analyzers, sensors, air conditioners, refrigerators, and semiconductor controls.

Partnership

NASA application of TE devices and relative technologies was influential in the products developed by Jim Kerner, chief executive officer and president of the Chico, California-based United States Thermoelectric Consortium Inc. (USTC). Since the debut of his first thermoelectric product programs in 1983, Kerner has built an outstanding international team of individuals who have designed and delivered significant TE device product solutions, and the company continues to grow in the arena of thermal management technologies implemented in military, aerospace, industrial, and consumer product areas.

Kerner has organized several companies in research, development, and manufacture of numerous thermoelectric-based products, including microprocessor validation tools, high-purity water delivery systems, portable refrigerator/warmer units, and precision temperature control (PTC) chambers. In 1988, *Spinoff* featured the water delivery system and a customized PTC chamber delivered to Ames Research Center, called the Portabator, which was manufactured in support of NASA Space Life Sciences Projects.

In cooperation with NASA Lewis Research Center, now Glenn Research Center, USTC built a gas emissions analyzer (GEA) for that Center's combustion research laboratory. As with the PTC product line, the GEA integrated TE devices in the design; a specification originated by NASA. This GEA precipitated hydrocarbon particles, preventing contamination that would hinder precise rocket fuel analysis. In addition to providing product development opportunities, affiliation with the NASA Industrial Application Center and Technology Transfer programs have been useful research tools for the company's subsequent product development programs.

USTC continues to integrate the benefits of TE devices in its current line of thermal management solutions and has found the accessibility of NASA technical research to be a valuable, sustainable resource that has continued to benefit and positively influence its product design and manufacturing.

United States Thermoelectric Consortium Inc. (USTC) has experienced great success in the development of thermal management devices based on micro-technologies, like this Rack Integrated Thermal Management System.

USTC is an experienced provider of fully integrated, turnkey thermal management solutions designed to meet stringent user requirements.

Product Outcome

TE devices are an important element in the precise temperature control of the USTC test validation tools for thermal management in the semiconductor and information technology industry. A USTC Thermal Tool System can precisely control the temperature of a heat-generated device up to 100 W/cm² in the range of -30 to 120 °C.

Among the USTC Thermal Tool System products are the Integrated Thermal Management System (ITMS), Hybrid Air Thermal Management System (HATMS), Rack Integrated Thermal Management System (RITMS), Temperature Control System (TCS), and Thermal Heads (TH). Integrating cooling modules and the TCS in one enclosure with liquid as working fluid, combined with TH, the ITMS executes temperature control tasks without other components. Using air as cooling media, the HATMS executes the tasks like ITMS. The RITMS includes a smart TCS unit which automatically adjusts parameters for temperature control, with an ultra-low noise level and low power consumption. The TCS is a standard temperature control unit, and the TH dissipate heat from a heat-generated device, like a computer. The RITMS or TCS, combined with a chiller (or a heat exchanger) and TH, provide a standard way to control the temperature and dissipate heat on electronic components for production validation tests.

The USTC research and design team uses patent-pending dimple, pin-fin, microchannel and microjet structures to develop and design heat dissipation devices on the mini-scale level, which not only guarantee high performance of products, but also scale device size from 1 centimeter to 10 centimeters.

The smart TCS unit in RITMS for temperature control simplifies the operation of the device. USTC provides choices for cooling media (liquid, air, and chilled liquid), high-power level, high-temperature range, fast-temperature response, scalable size, remote control, low noise, highly reliable, low maintenance, and easy operation. USTC also creates, develops, and delivers solutions for custom thermal management requirements.

The USTC team has been constantly innovating with the latest relevant technologies to provide the best thermal solutions for computer, processor, chipset, board, drawer, server, and rack challenges, as well as other thermal management areas like laser, microwave, radar, backup energy storage, and thermo-stabilization systems.

Since 1997, USTC solutions have been integrated into microprocessor tests, validation processes, and design debugging, and have been deployed to all major development centers in the United States and abroad, including to China, the United Kingdom, Israel, India, Malaysia, and the Philippines.

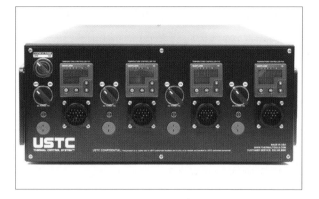

USTC has received over 30 U.S.- and foreign-issued patents and has patent applications pending on new methods and devices.

Kerner organized USTC with a vision to build a world-class team of experts to deal with Moore's Law: as processor speed increases, power and heat will also increase, and at a certain point, processors cannot continue to operate with increased speeds unless thermals are managed. A unique team of scientists and engineers are organized to focus on research and design, as well as the application of new technologies in thermal management.

The USTC team includes 15 Ph.D. scientists and other specialists with core expertise in thermal management, amplified by a wide spectrum of experience in areas including thermophysics, aero and fluid dynamics, porous materials, capillary structures, microchannels, and refrigeration systems. USTC has received over 30 U.S.- and foreign-issued patents and pending patent applications on new methods and devices. ❖

Research and Operations

Pushing back boundaries in aeronautics and space exploration relies upon the ongoing research activities and operational support led by the four Mission Directorates: Science, Exploration Systems, Aeronautics Research, and Space Operations. These efforts are conducted at each of NASA's 10 field centers.

Research and Operations

NASA's four Mission Directorates support the Agency's missions to resume the human exploration of the Moon and onward to Mars.

- The Science Mission Directorate engages the Nation's science community, sponsors scientific research, and develops and deploys satellites and probes in collaboration with NASA's partners around the world to answer fundamental questions requiring the view from and into space.

- The Exploration Systems Mission Directorate develops the systems that will enable NASA to embark on a robust space exploration program that will advance the Nation's scientific, security, and economic interests.

- The Aeronautics Research Mission Directorate cements NASA's role as the leading government organization for aeronautical research, with world-class capability built on a tradition of expertise in aeronautical engineering and its core research areas, including aerodynamics, aeroacoustics, materials and structures, propulsion, dynamics and control, sensor and actuator technologies, advanced computational and mathematical techniques, and experimental measurement techniques.

- The Space Operations Mission Directorate provides leadership and management of NASA space operations related to human exploration in and beyond low-Earth orbit. This includes the Space Shuttle and International Space Station programs. The directorate is also responsible for launch services and space communications in support of both human and robotic exploration.

Science Mission Directorate

The Science Mission Directorate (SMD) seeks to understand the origins, evolution, and destiny of the universe, and the phenomena that shape it. SMD also works to increase our understanding of the solar system, the Sun, and the Earth, and the nature of life in the universe and what kinds of life may exist beyond Earth.

NASA's Cassini Discovers Potential Liquid Water on Enceladus

NASA's Cassini spacecraft may have found evidence of liquid-water reservoirs that erupt in Yellowstone-like geysers on Saturn's moon Enceladus. The rare occurrence of liquid water so near the surface raises many new questions about the mysterious moon.

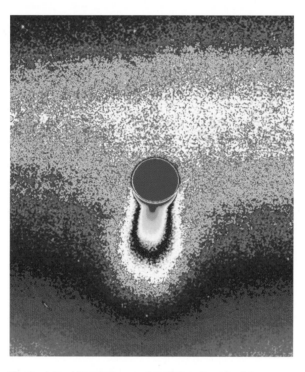

The ice jets of Enceladus send particles streaming into space hundreds of kilometers above the south pole of this spectacularly active moon. Some of the particles escape to form the diffuse E ring around Saturn. This color-coded image was processed to enhance faint signals, making the contours and extent of the fainter, larger-scale component of the plume easier to see.

High-resolution Cassini images showed icy jets and towering plumes ejecting large quantities of particles at high speed. Scientists examined several models to explain the process. They ruled out the idea the particles are produced or blown off the surface by vapor created when warm water ice converts to a gas. Instead, scientists have found evidence for a much more exciting possibility. The jets might be erupting from near-surface pockets of liquid water above 32 °F, like cold versions of the Old Faithful geyser in Yellowstone National Park.

Other moons in the solar system have liquid-water oceans covered by kilometers of icy crust. What's different here is that pockets of liquid water may be no more than tens of meters below the surface.

Scientists still have many questions. Why is Enceladus so active? Are other sites on Enceladus active? Might this activity have been continuous enough over the moon's history for life to have had a chance to take hold in the moon's interior? In the spring of 2008, scientists will get another chance to look at Enceladus when Cassini flies within 350 kilometers (approximately 220 miles).

The Cassini-Huygens mission is a cooperative project of NASA, the European Space Agency, and the Italian Space Agency. NASA's Jet Propulsion Laboratory (JPL), a division of the California Institute of Technology, manages the Cassini-Huygens mission for NASA's Science Mission Directorate. The Cassini orbiter was designed, developed, and assembled at JPL.

NASA Images Suggest Water Still Flows in Brief Spurts on Mars

NASA photographs revealed bright new deposits seen in two gullies on Mars that suggest water carried sediment through them sometime during the past 6 years. These observations give the strongest evidence to date that water still flows occasionally on the surface of Mars.

Liquid water, as opposed to the water ice and water vapor known to exist on Mars, is considered necessary for life. The new findings heighten intrigue about the poten-

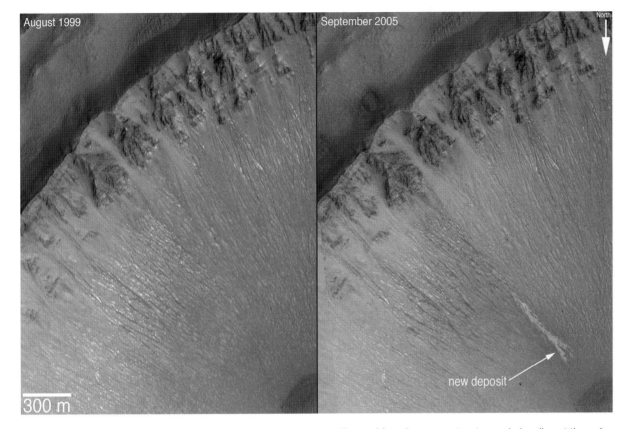

August 1999   September 2005   North

300 m

new deposit

NASA photographs have revealed bright new deposits seen in two gullies on Mars that suggest water carried sediment through them sometime during the past 6 years.

tial for microbial life on Mars. The Mars Orbiter Camera on NASA's Mars Global Surveyor provided the new evidence of the deposits in images taken in 2004 and 2005.

"The shapes of these deposits are what you would expect to see if the material were carried by flowing water," said Michael Malin of Malin Space Science Systems, San Diego. "They have finger-like branches at the downhill end and easily diverted around small obstacles." Malin is principal investigator for the camera and lead author of a report about the findings published in the journal Science.

The atmosphere of Mars is so thin and the temperature so cold that liquid water cannot persist at the surface. It would rapidly evaporate or freeze. Researchers propose that water could remain liquid long enough, after breaking out from an underground source, to carry debris downslope before totally freezing. The two fresh deposits are each several hundred meters long.

The light tone of the deposits could be from surface frost continuously replenished by ice within the body of the deposit. Another possibility is a salty crust, which would be a sign of water's effects in concentrating the salts. If the deposits had resulted from dry dust slipping down the slope, they would likely be dark, based on the dark tones of dust freshly disturbed by rover tracks, dust devils, and fresh craters on Mars.

The Mars Global Surveyor has discovered tens of thousands of gullies on slopes inside craters and other depressions on Mars. Most gullies are at latitudes of 30° or higher. Malin and his team first reported the discovery of the gullies in 2000. To look for changes that might indicate present-day flow of water, his camera team repeatedly imaged hundreds of the sites. One pair of images showed a gully that appeared after mid-2002. That site was on a sand dune, and the gully-cutting process was interpreted as a dry flow of sand.

The announcement is the first to reveal newly deposited material apparently carried by fluids after earlier imaging of the same gullies. The two sites are inside craters in the Terra Sirenum and the Centauri Montes regions of southern Mars.

"These fresh deposits suggest that at some places and times on present-day Mars, liquid water is emerging from beneath the ground and briefly flowing down the slopes. This possibility raises questions about how the water would stay melted below ground, how widespread it might be, and whether there's a below-ground wet habitat conducive to life. Future missions may provide the answers," said Malin.

Besides looking for changes in gullies, the orbiter's camera team assessed the rate at which new impact craters appear. The camera photographed approximately 98 percent of Mars in 1999 and approximately 30 percent of the planet was photographed again in 2006. The newer images show 20 fresh impact craters ranging in diameter from 7 feet to 486 feet that were not present approximately 7 years earlier. These results have important implications for determining the ages of features on the surface of Mars. These results also approximately match predictions and imply that Martian terrain with few craters is truly young.

Mars Global Surveyor began orbiting Mars in 1997. The spacecraft is responsible for many important discoveries. NASA declared early in 2007 that the spacecraft is no longer operating.

Pluto-Bound New Horizons Provides New Look at Jupiter System

NASA's New Horizons spacecraft provided new data on the Jupiter system, stunning scientists with never-before-seen perspectives of the giant planet's atmosphere, rings, moons, and magnetosphere.

These new views include the closest look yet at the Little Red Spot storm churning materials through Jupiter's cloud tops; detailed images of small satellites herding dust and boulders through Jupiter's faint rings; and volcanic eruptions and circular grooves on the planet's largest moons.

New Horizons came to within 1.4 million miles of Jupiter on February 28, 2007, using the planet's gravity to trim 3 years from its travel time to Pluto. For several weeks before and after this closest approach, the piano-sized robotic probe trained its 7 cameras and sensors on Jupiter and its 4 largest moons, storing data from nearly 700 observations on its digital recorders and gradually sending that information back to Earth. Nearly all of the expected 34 gigabits of data has come back so far, radioed to NASA's largest antennas over more than 600 million miles. This activity confirmed the successful testing of the instruments and operating software the spacecraft will use at Pluto.

Aside from setting up the 2015 arrival at Pluto, the Jupiter flyby was a stress test of NASA's spacecraft and team, and both passed with very high marks.

Images include the first close-up scans of the Little Red Spot, Jupiter's second-largest storm, which formed when three smaller storms merged during the past decade. The storm, about half the size of Jupiter's larger Great Red Spot and about 70 percent of Earth's diameter, began turning red about a year before New Horizons flew

Image of the planet Jupiter's moon, Io, as seen by the New Horizons spacecraft. A plume from a huge volcanic eruption can be seen at the north pole.

past it. Scientists will search for clues about how these systems form and why they change colors in their close observations of materials spinning within and around the nascent storm.

Under a range of lighting and viewing angles, New Horizons also grabbed the clearest images ever of the tenuous Jovian ring system. In them, scientists spotted a series of unexpected arcs and clumps of dust, indicative of a recent impact into the ring by a small object. Movies made from New Horizons images also provide an unprecedented look at ring dynamics, with the tiny inner moons Metis and Adrastea appearing to shepherd the materials around the rings.

Of Jupiter's four largest moons, the team focused much attention on volcanic Io, the most geologically active body in the solar system. New Horizons' cameras captured pockets of bright, glowing lava scattered across the surface; dozens of small, glowing spots of gas; and several fortuitous views of a sunlit umbrella-shaped dust plume rising 200 miles into space from the volcano Tvashtar, the best images yet of a giant eruption from the tortured volcanic moon.

The timing and location of the spacecraft's trajectory also allowed it to spy many of the mysterious, circular troughs carved onto the icy moon Europa. Data on the size, depth, and distribution of these troughs, discovered by the Jupiter-orbiting Galileo mission, will help scientists determine the thickness of the ice shell that covers Europa's global ocean.

Already the fastest spacecraft ever launched, New Horizons reached Jupiter 13 months after lifting off from Cape Canaveral Air Force Station, in January 2006. The flyby added 9,000 miles per hour, pushing New Horizons past 50,000 miles per hour and setting up a Pluto flyby in July 2015.

The number of observations at Jupiter was twice that of those planned at Pluto. New Horizons made most of these observations during the spacecraft's closest approach to the planet, which was guided by more than 40,000 separate commands in the onboard computer.

New Horizons is the first mission in NASA's New Frontiers Program of medium-class spacecraft exploration projects. Alan Stern, the Science Mission Directorate associate administrator and New Horizons principal investigator, leads the mission and science team; the Johns Hopkins University Applied Physics Laboratory manages the mission for NASA's Science Mission Directorate. The mission team also includes Ball Aerospace and Technologies Corporation, The Boeing Company, KinetX Inc., Lockheed Martin Corporation, Stanford University, University of Colorado at Boulder, the U.S. Department of Energy, and a number of other firms, university partners, and NASA centers, including JPL and Goddard Space Flight Center.

Dark matter and normal matter have been wrenched apart by the tremendous collision of two large clusters of galaxies. The discovery, using NASA's Chandra X-ray Observatory and other telescopes, gives direct evidence for the existence of dark matter.

NASA Finds Direct Proof of Dark Matter

Dark matter and normal matter have been wrenched apart by the tremendous collision of two large clusters of galaxies. The discovery, using NASA's Chandra X-ray Observatory and other telescopes, gives direct evidence for the existence of dark matter.

This is the most energetic cosmic event, besides the Big Bang, that we know about. These observations provided the strongest evidence yet that most of the matter in the universe is dark. Despite considerable evidence for dark matter, some scientists have proposed alternative theories for gravity where it is stronger on intergalactic scales than predicted by Newton and Einstein, removing the need for dark matter. However, such theories cannot explain the observed effects of this collision.

In galaxy clusters, the normal matter, like the atoms that make up the stars, planets, and everything on Earth, is primarily in the form of hot gas and stars. The mass of the hot gas between the galaxies is far greater than the mass of the stars in all of the galaxies. This normal matter is bound in the cluster by the gravity of an even greater mass of dark matter. Without dark matter, which is invisible and can only be detected through its gravity, the fast-moving galaxies and the hot gas would quickly fly apart.

The team that made this discovery was granted more than 100 hours on the Chandra telescope to observe the galaxy cluster 1E0657-56. The cluster is also known as the bullet cluster, because it contains a spectacular bullet-shaped cloud of 100-million-degree gas. The X-ray image shows the bullet shape is due to a wind produced by the high-speed collision of a smaller cluster with a larger one.

In addition to the Chandra observation, the Hubble Space Telescope, the European Organization for Astronomical Research in the Southern Hemisphere's Very Large Telescope, and the Magellan optical telescopes were used to determine the location of the mass in the clusters. This was done by measuring the effect of gravitational lensing, where gravity from the clusters distorts light from background galaxies as predicted by Einstein's theory of general relativity.

The hot gas in this collision was slowed by a drag force, similar to air resistance. In contrast, the dark matter was not slowed by the impact, because it does not interact directly with itself or the gas except through gravity. This produced the separation of the dark and normal matter seen in the data. If hot gas was the most massive component in the clusters, as proposed by alternative gravity theories, such a separation would not have been seen. Instead, dark matter is required.

This is the type of result that future theories will have to take into account. As we move forward to understand the true nature of dark matter, this new result will be impossible to ignore. It also gives scientists more confidence that the Newtonian gravity, familiar on Earth and in the solar system, also works on the huge scales of galaxy clusters.

NASA Research Reveals Climate Warming Reduces Ocean Food Supply

In a recent NASA study, scientists concluded that when Earth's climate warms, there is a reduction in the ocean's primary food supply. This poses a potential threat to fisheries and ecosystems. By comparing nearly a decade of global ocean satellite data with several records of Earth's changing climate, scientists found that whenever climate temperatures warmed, marine plant life in the form of microscopic phytoplankton declined. Whenever climate temperatures cooled, marine plant life became more vigorous or productive. The results provide a preview of what could happen to ocean biology in the future if Earth's climate warms as the result of increasing levels of greenhouse gasses in the atmosphere.

Phytoplankton is composed of microscopic plants living in the upper sunlit layer of the ocean. It is responsible for approximately the same amount of photosynthesis each year as all land plants combined. Changes in phytoplankton growth and photosynthesis influence fishery yields, marine bird populations, and the amount of carbon dioxide the oceans remove from the atmosphere.

"Rising levels of carbon dioxide in the atmosphere play a big part in global warming," said lead author Michael Behrenfeld, of Oregon State University, Corvallis. "This study shows that as the climate warms, phytoplankton growth rates go down and along with them the amount of carbon dioxide these ocean plants consume. That allows carbon dioxide to accumulate more rapidly in the atmosphere, which would produce more warming."

The findings were from a NASA-funded analysis of data from the Sea-viewing Wide Field-of-view Sensor (SeaWiFS) instrument on the OrbView-2 spacecraft, launched in 1997. SeaWiFS is jointly operated by GeoEYE, of Dulles, Virginia, and NASA.

The uninterrupted 9-year record shows in great detail the ups and downs of marine biological activity or productivity from month to month and year to year. Captured at the start of this data record was a major, rapid rebound in ocean biological activity after a major El Niño event. El Niño and La Niña are major warming and cooling events, respectively, that occur approximately every 3 to 7 years in the eastern Pacific Ocean and are known to change weather patterns around the world.

Scientists made their discovery by comparing the SeaWiFS record of the rise and fall of global ocean plant life to different measures of recent global climate change. The climate records included several factors that directly affect ocean conditions, such as changes in sea surface temperature and surface winds. The results support computer model predictions of what could happen to the world's oceans as the result of prolonged future climate warming.

Ocean plant growth increased from 1997 to 1999 as the climate cooled during one of the strongest El Niño to La Niña transitions on record. Since 1999, the climate has been in a period of warming that has seen the health of ocean plants diminish.

The new study also explains why a change in climate produces this effect on ocean plant life. When the climate warms, the temperature of the upper ocean also increases, making it "lighter" than the denser cold water beneath it. This results in a layering or "stratification" of ocean waters that creates an effective barrier between the surface layer and the nutrients below, cutting off phytoplankton's food

Scientists now have nearly a decade's worth of data showing the cycle of plant life in the Earth's oceans. From space, the "ocean color" satellites measure the ocean's biology as plant productivity. In this visualization, high plant productivity is represented in green, while areas of low productivity remain blue.

supply. The scientists confirmed this effect by comparing records of ocean surface water density with the SeaWiFS biological data.

NASA Spacecraft Make First 3-D Images of Sun

NASA's twin Solar Terrestrial Relations Observatory (STEREO) spacecraft made the first three-dimensional images of the Sun. The new view will greatly aid scientists' ability to understand solar physics and thereby improve space weather forecasting.

"The improvement with STEREO's 3-D view is like going from a regular X-ray to a 3-D CT [computerized tomography] scan in the medical field," said Dr. Michael Kaiser, STEREO project scientist at Goddard.

The STEREO spacecraft were launched October 25, 2006. On January 21, 2007, they completed a series of complex maneuvers, including flying by the Moon, to position the spacecraft in their mission orbits. The two observatories are now orbiting the Sun, one slightly ahead of Earth and one slightly behind, separating from each other by approximately 45° per year. Just as the slight offset between a person's eyes provides depth perception, the separation of these spacecraft allow 3-D images of the Sun.

Violent solar weather originates in the Sun's atmosphere, or corona, and can disrupt satellites, radio communication, and power grids on Earth. The corona resembles wispy smoke plumes, which flow outward along the Sun's tangled magnetic fields. It is difficult for scientists to tell which structures are in front and which are behind.

With STEREO's 3-D imagery, scientists will be able to discern where matter and energy flows in the solar atmosphere much more precisely than with the 2-D views available before.

STEREO's depth perception also will help improve space weather forecasts. Of particular concern is a destructive type of solar eruption called a coronal mass ejection

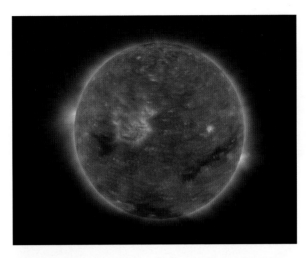

NASA's STEREO satellites have provided the first 3-D images of the Sun.

(CME). CMEs are eruptions of electrically charged gas, called plasma, from the Sun's atmosphere. A CME cloud can contain billions of tons of plasma and move at a million miles per hour.

The CME cloud is laced with magnetic fields, and CMEs directed toward Earth smash into its magnetic field. If the CME magnetic fields have the proper orientation, they dump energy and particles into Earth's magnetic field, causing magnetic storms that can overload power line equipment and radiation storms that disrupt satellites.

Satellite and utility operators can take precautions to minimize CME damage, but they need an accurate forecast of when the CME will arrive. To do this, forecasters need to know the location of the front of the CME cloud. STEREO will allow scientists to accurately determine this location. Knowing where the front of the CME cloud is will improve estimates of the arrival time from within a day or so to just a few hours. STEREO also will help forecasters estimate how severe the resulting magnetic storm will be.

"In addition to the STEREO perspective of solar features, STEREO, for the first time, will allow imaging of the solar disturbances the entire way from the Sun to the Earth. Presently, scientists are only able to model this region in the dark, from only one picture of solar disturbances leaving the Sun and reaching only a fraction of the Sun-Earth distance," said Dr. Madhulika Guhathakurta, a STEREO program scientist at NASA Headquarters.

STEREO's first 3-D images are being provided by JPL. STEREO is the third mission in NASA's Solar Terrestrial Probes program within NASA's Science Mission Directorate. The Goddard Science and Exploration Directorate manages the mission, instruments, and science center. The Johns Hopkins University Applied Physics Laboratory designed and built the spacecraft and is responsible for mission operations. The STEREO imaging and particle detecting instruments were designed and built by scientific institutions in the United States, United Kingdom, France, Germany, Belgium, Netherlands, and Switzerland.

Pioneering NASA Spacecraft Mark 30 Years of Flight

NASA's two venerable Voyager spacecraft are celebrating 3 decades of flight as they head toward interstellar space; Voyager 2 launched on August 20, 1977, and Voyager 1 launched on September 5, 1977. Their ongoing odysseys mark an unprecedented and historic accomplishment, as they continue to return information from distances more than 3 times farther away than Pluto.

Voyager 1 is currently the farthest-traveled human-made object, at a distance of about 9.7 billion miles (15.5 billion kilometers) from the Sun. Voyager 2 is about 7.8 billion miles (12.5 billion kilometers) from the Sun. Originally designed as a 4-year mission to Jupiter and Saturn, the Voyager tours were extended because of their successful achievements and a rare planetary alignment. The two-planet mission eventually became a four-planet grand tour, and after completing that

Voyager 1 passed the Saturnian system in November 1980; 9 months later, Voyager 2 passed through this same system. The ensuing scientific discoveries were unprecedented with regards to the rings around Saturn and its satellite's chemical makeup. Pictured are: Saturn (shown with rings), Dione (forefront), Tethys and Mimas (lower right), Enceladus and Rhea (upper left) and Titan in distant orbit (upper right).

extended mission, the two spacecraft began the task of exploring the outer heliosphere.

During their first dozen years of flight, the Voyagers made detailed explorations of Jupiter, Saturn, and their moons, and conducted the first explorations of Uranus and Neptune. The Voyagers returned never-before-seen images and scientific data, making fundamental discoveries about the outer planets and their moons. The spacecraft revealed Jupiter's turbulent atmosphere, which includes dozens of interacting hurricane-like storm systems, and erupting volcanoes on Jupiter's moon, Io. They also showed waves and fine structure in Saturn's icy rings from the tug of nearby moons.

"The Voyager mission has opened up our solar system in a way not possible before the Space Age," said Edward Stone, Voyager project scientist at the California Institute of Technology. "It revealed our neighbors in the outer solar system and showed us how much there is to learn and how diverse the bodies are that share the solar system with our own planet Earth."

In December 2004, Voyager 1 ventured into the solar system's final frontier. Called the heliosheath, this turbulent area, approximately 8.7 billion miles (14 billion kilometers) from the Sun, is where the solar wind slows as it crashes into the thin gas that fills the space between stars. Voyager 2 could reach this boundary in late 2007, putting both Voyagers on their final leg toward interstellar space.

Each Voyager logs approximately 1 million miles per day and carries five science instruments that study the solar wind, energetic particles, magnetic fields, and radio waves as they cruise through this unexplored region of deep space. While the spacecraft are now too far from the Sun to use solar power, their long-lived radioisotope thermoelectric generators provide the necessary power of less than 300 watts (the amount of power needed to light up a bright light bulb). The Voyagers call home via NASA's Deep Space Network, a system of antennas around the world. The spacecraft are so distant that commands from Earth, traveling at light speed, take 14 hours one-way to reach Voyager 1 and 12 hours to reach Voyager 2.

"The continued operation of these spacecraft and the flow of data to the scientists is a testament to the skills and dedication of the small operations team," said Ed Massey, Voyager project manager at NASA's Jet Propulsion Laboratory. Massey oversees a team of nearly a dozen people in the day-to-day Voyager spacecraft operations.

Each of the Voyagers carries a golden record that is a time capsule with greetings, images, and sounds from Earth. The records also have directions on how to find Earth should the spacecraft be recovered.

NASA's latest outer planet exploration mission is New Horizons, which is now well past Jupiter and headed for a historic exploration of the Pluto system in July 2015.

Exploration Systems Mission Directorate

The Exploration Systems Mission Directorate develops the launch systems, vehicles, and other capabilities that will carry humans into space and ultimately enable exploration of the Moon and Mars, including the servicing of the International Space Station following the retirement of the space shuttle in 2010.

Global Exploration Strategy

Driven by the Vision for Space Exploration and guided by the NASA Authorization Act of 2005, the United States will take the next steps in human space exploration with thoughtful preparation and global partnership. NASA is tackling the challenging questions of how, why, and when the United States will return to the Moon. To date, NASA has worked with more than 1,000 people around the world to craft an initial strategy. NASA has sought perspectives on exploration strategy from 13 international space agencies, as well as U.S. entities including the space industry, academia, and other government organizations. The Agency issued a request for information and held international meetings, teleconferences, and bilateral exchanges. From the process, NASA characterized 6 strategic themes and prioritized 188 objectives that were synthesized from 800 potential strategic objectives.

The strategic themes and objectives represent the driving force behind NASA's current plans for human and robotic exploration on the lunar surface. The strategic themes substantiate the answers to the question, "Why are we going to the Moon?" The objectives address the question, "What do we hope to accomplish when we get there?" The six themes are:

- Exploration Preparation: Prepare for future human and robotic missions to Mars and other destinations.

- Scientific Knowledge: Pursue scientific activities addressing fundamental questions about Earth, the solar system, the universe, and our place in them.

- Human Civilization: Extend human presence in space.

- Economic Expansion: Expand Earth's economic sphere and conduct activities with benefits to life on Earth.

- Global Partnerships: Strengthen existing partnerships and create new ones.

- Public Engagement: Engage, inspire, and educate the public.

From these 6 themes, NASA derived 25 categories of lunar objectives. The categories are: astronomy and astrophysics; commerce; commercial opportunities; communication; crew activity support; Earth observation; environmental characterization; environmental hazard mitigation; general infrastructure; geology; global partnership; guidance, navigation, and control; heliophysics; historic preservation; human health; life support and habitat; lunar resource utilization; materials science; operational environmental monitoring; operations test and verification; power; program execution; public engagement and inspiration; surface mobility; and transportation.

Lunar Architecture Plan

Concurrently in 2006, NASA chartered the Lunar Architecture Team, to play a key role in the continual development and implementation of the global exploration strategy. The comprehensive effort includes contributions from more than 200 representatives from

This artist's concept shows a lunar lander docked with an Orion crew vehicle while in orbit around the Moon. The Orion capsule will remain empty while the astronauts descend in the lunar lander to explore the Moon's surface.

NASA field centers. The Lunar Architecture Team charter directs the team to accomplish four key tasks: 1) Develop a baseline architecture for robotic and human lunar missions that can be traced directly to specific objectives; 2) formulate a concept of operations for planned lunar missions; 3) develop individual requirements that will be incorporated into NASA's exploration architecture requirements document; and 4) assess functional needs

A lander could transport explorers and a scientific payload to other lunar sites.

Incremental build up of an outpost would begin with the first mission of the human lunar campaign and each subsequent mission would add to the useful infrastructure.

and analyze required technologies. The architecture represents NASA's current plan for returning to the Moon using both robotic and human missions. The architecture will evolve through future studies and further refinement of the requirements.

Through comprehensive and thorough analysis, the Lunar Architecture Team drafted a baseline lunar architecture rooted in the strategic themes and objectives as articulated in the team's charter. The document presents a summary of the baseline architecture. Phase 2 Lunar Architecture Team activities in 2007 include more specific lunar architecture element trade studies and refinement of level one, or fundamental, requirements.

The Lunar Architecture Team analyzed two basic approaches to human lunar exploration. One was the lunar sortie, which employs distinct short-duration human missions to one or more lunar locations before any build up of a permanent outpost. The team concluded that building an outpost first will better address the entire portfolio of strategic themes and objectives and enable many scientific objectives.

The "outpost first" approach focuses on a single lunar site. Incremental build up of an outpost would begin with the first mission of the human lunar campaign, and each subsequent mission would add to the useful infrastructure. In a completed outpost, crews could live and work on the Moon in 6-month rotations. They also could make frequent and increasingly distant trips away from the outpost to explore, conduct field experiments, and analyze scientific data with a high level of comfort and safety.

In keeping with the strategic themes and objectives, a lunar outpost would extend human civilization and serve as a test bed for future missions to Mars and other destinations. Thoughtful site selection could help astronauts on future missions utilize local resources, a potentially critical component of human missions to Mars. An outpost also enables many science-oriented activities. A Lunar Architecture Team review of about 72 science objectives concluded that more than half could be substantially accomplished within 7 years of initiating the planned architecture. When outpost-based science and robotic sorties are added to the current architecture, more than 90 percent of the rated scientific objectives identified by the global strategy process could be addressed.

Construction of an outpost would not preclude sortie missions to other lunar locations. A lander with extra fuel cells could transport two explorers and a substantial science payload to other lunar sites.

The Moon may harbor natural resources that could enable crews to "live off the land" and even prepare for trips to Mars and other destinations. The rocky lunar surface layer, called regolith, is a potential source of oxygen, for instance. The poles, in particular, offer concentrations of hydrogen and possibly water ice.

In considering where to situate an outpost, the Lunar Architecture Team determined that a polar location would be an accessible, simple place to operate and offered advantages such as natural resources and the potential for interesting science. The poles are less constrained by lighting and orbital phasing considerations, providing more launch opportunities. Certain polar locations are more temperature-stable and have shorter periods of darkness, reducing power and thermal system requirements. Use of solar power would allow explorers to quickly and inexpensively develop the ability for extended stays. Polar locations offer great potential for meeting science objectives. We know less about the poles than other areas of the Moon, and they offer the unique feature of cold, dark craters.

According to the Lunar Architecture Team, the best understood area near the South Pole of the Moon is near the rim of the Shackleton crater. The area is roughly the same size as the National Mall in Washington, D.C. It is sunlit about 80 percent of the time during the southern lunar winter and even longer during the southern lunar summer. An outpost near the edge offers access to an area of permanent darkness inside the crater, a chance to meet some top science objectives, and the possibility of a rich source of water ice. For all these reasons, Shackleton crater serves as a departure point for planning purposes. NASA will select a final outpost location later, based on analyses of scientific data gathered by robot probes.

In order to provide sustained human presence and serve as a stepping stone for the future exploration of Mars, the lunar outpost must employ long-duration systems and enable autonomous and robust surface operations. Routine exploration beyond the outpost site, reliable communication, teleoperation and system autonomy, and reliance on live-off-the-land resources will be crucial. The architecture allows astronauts to return significant quantities of lunar samples. The outpost design will accommodate sites other than the poles.

It will take a significant effort over many years to build up the complete lunar outpost where humans can live and work on a continuous basis.

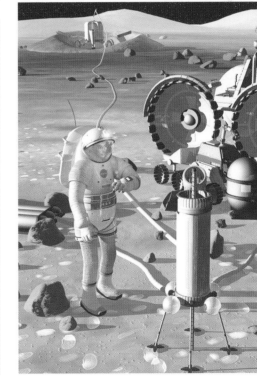

On the Moon, NASA astronauts will pursue a variety of scientific activities.

The LRO is the first in a series of missions to the Moon, planned for launch in late 2008 and orbiting for at least 1 year.

In preparation for human exploration of Mars, activities at the lunar outpost must enable a better understanding of the long-term physiological effects of living and working on another planetary body. Astronauts will stay healthy and productive through the use of preventive medicine; telemedicine and trauma care; exercise regimens and nutrition countermeasures; radiation protection and mitigation; and research on bone loss, cardiovascular and cardiopulmonary function, musculoskeletal status, and neurological function.

Within the lunar architecture, NASA will pursue scientific activities corresponding to the global exploration strategic theme that addresses fundamental questions about Earth, the solar system, the universe, and our place in them. Potential geoscience activities include investigation of the layered nature and formation process of the lunar regolith and study of lunar volatiles. Field operations will include use of teleoperated robotic explorers and the ability to perform preliminary chemical and mineralogical analyses on geologic samples. The geoscience activities performed on the Moon will help provide the equipment and techniques that will enable the efficient geologic exploration of Mars and other destinations in the solar system. In addition, the knowledge gained on the basic geologic processes, such as meteoritic impact and volcanism, will help scientists better understand how these processes occur throughout the solar system. NASA can also pursue other science such as space physics and astronomy under this architecture. The science that is pursued will be under the purview and selection processes of NASA's Science Mission Directorate.

In the spirit of global participation, NASA has adopted an open architecture approach to lunar surface infrastructure and activities that encourage external involvement (such as use of the metric system).

NASA also seeks to conduct lunar exploration in a way that engages, inspires, and educates the general public. The architecture includes allocations for payloads specifically designed to support public interest, including a network of cameras. Non-governmental organizations will be able to control some of these cameras.

It will take a significant effort over many years to build up the complete lunar outpost where humans can live and work on a continuous basis. In fact, much work remains to be done before astronauts can even set foot on the lunar surface. Years before humans can successfully return to the surface of the Moon, the way must be prepared. The Lunar Reconnaissance Orbiter (LRO) and the Lunar Crater Observation and Sensing Satellite (LCROSS) will collect essential information about the future site of the outpost. The LRO and the LCROSS will launch together from the Kennedy Space Center in late 2008. The LRO will orbit the Moon for at least one year, creating high-resolution maps of the lunar terrain, seeking ideal landing sites by identifying hazards, and characterizing the thermal, lighting, and radiation environment of the Moon.

Mission planners intend for the LCROSS to impact the surface of the Moon in a crater at one of the poles in an attempt to measure water that may be frozen in the lunar soil. The water-seeking satellite and the LRO's Earth departure stage booster will strike the Moon's South Pole in January 2009. As the orbiter approaches the Moon, the Earth departure stage will separate and impact a crater, creating a 2.2-million-pound plume. The satellite, a shepherding spacecraft, will fly through the plume, using instrumentation to examine the cloud for signs of water and other compounds. At the end of its mission, the satellite itself will become an impactor, creating a second plume visible to lunar-orbiting spacecraft and Earth-based observatories.

Lunar Launch Strategies

NASA plans to return humans to the Moon no later than 2020. This journey has already begun, with the development of a new spaceship. Building on the best of past and present technology, NASA is creating a 21st-century exploration system that will be affordable, reliable, versatile, and safe.

The centerpiece of this system is the new spacecraft, Orion, designed to carry four astronauts to and from the Moon, deliver up to six crew members and supplies to the International Space Station, and support future missions to Mars. Coupled with a new lunar lander, Orion will send twice as many astronauts to the lunar surface as Apollo and allow them to stay longer, with initial missions lasting 4 to 7 days. While Apollo was limited to landings along the Moon's equator, this lunar lander will carry enough propellant to land anywhere on the Moon's surface.

Once a lunar outpost is established, crews could remain on the surface for up to 6 months. Orion will autonomously operate for up to 6 months in lunar orbit, standing by to return human explorers to Earth.

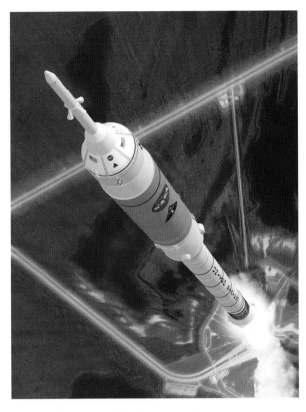

A concept image shows the Ares I crew launch vehicle during ascent. Ares I is an in-line, two-stage rocket configuration topped by the Orion crew exploration vehicle and launch abort system. The Ares I first stage is a single, five-segment reusable solid rocket booster, derived from the space shuttle. Its upper stage is powered by a J-2X engine. Ares I will carry the Orion with its crews of up to six astronauts to Earth orbit.

The launch system that will get the crew off the ground builds on powerful, reliable propulsion elements. Astronauts will launch on the Ares I rocket made up of a five-segment shuttle solid rocket booster, with a second stage powered by a J-2X engine, based on engines used on the Apollo Saturn V rockets.

A second, heavy-lift rocket, the Ares V, will use the five-segment solid rocket boosters and five liquid-fueled RS-68 engines to put up to 125 metric tons in orbit—slightly more than the weight of a shuttle orbiter. This versatile system will be capable of supporting lunar and Mars missions.

Aeronautics Research Mission Directorate

The Aeronautics Research Mission Directorate (ARMD) generates the revolutionary concepts, technologies, and capabilities needed to advance aircraft and airspace systems. ARMD's programs facilitate safer, more efficient, and more environmentally friendly air transportation systems. In addition, ARMD's research will continue to play a vital role in NASA's human and robotic space activities.

Aeronautics Overview

NASA has been at the forefront of aeronautics research for more than 9 decades. In 1915, the United States faced a critical need to learn everything it could, as fast as it could, about the science of flight. In response to this need, President Woodrow Wilson created the National Advisory Committee for Aeronautics (NACA). Through two World Wars, NACA research contributed to the design and development of every American aircraft in both the commercial and military sector. In 1958, NACA's aeronautics work was transferred to the newly formed National Aeronautics and Space Administration, NASA.

Today, the ARMD efforts are directed toward the transformation of the Nation's air transportation system, and developing the knowledge, tools, and technologies to support future air and space vehicles. The three major programs within ARMD that carry out this research are the Fundamental Aeronautics Program, the Aviation Safety Program, and the Airspace Systems Program.

ARMD's focus is on cutting-edge, foundational research in traditional aeronautical disciplines, as well as in emerging fields with promising application to aeronautics. ARMD is investing in research for the long-term in areas that are appropriate to NASA's unique capabilities, and to meeting its charter of addressing national needs and benefiting the public good. The Directorate is advancing the science of aeronautics as a resource to the Nation, as well as advancing technologies, tools, and system concepts that can be drawn upon by civilian and military communities and other government agencies.

Being committed to developing tools and technologies can help transform how air transportation systems operate, how new aircraft are designed and manufactured, and how the Nation's air transportation system can reach unparalleled levels of safety, capacity, and efficiency.

Advanced Composites

Recent efforts under NASA's Small Business Innovative Research (SBIR) program have shown the technical feasibility and potential benefits of advanced composite materials/structures technologies in demanding aircraft applications. NASA's Glenn Research Center and its industry and university partners have demonstrated the feasibility of two advanced composite technologies for future jet engine fan casing/containment system applications with the potential for 25- to 40-percent weight reductions and equal or better performance as measured by structural and damage tolerance parameters.

The first technology, which uses fiber preform braiding and resin transfer molding processes, has been demonstrated in subsystem and engine level testing and is on a path for commercial engine certification by two engine manufacturers, with entry into commercial service expected to occur in the next year or two. This technology has been developed by A&P Technology Inc., under NASA SBIR Phase I, II, and III contracts and also with additional investment from ARMD's Aviation Safety Program.

The second technology, which uses fiber-reinforced-foam core preform winding/stitching and resin vacuum infusion molding processes, is in an earlier stage of development but has been demonstrated in lab and rig

A jet engine fan casing constructed of advanced composite technologies created by Glenn Research Center and its industry and university partners.

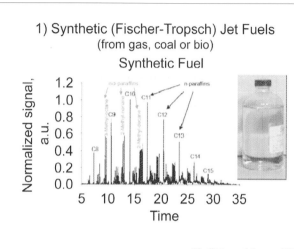

1) Synthetic (Fischer-Tropsch) Jet Fuels
(from gas, coal or bio)

2) Bioderived Renewable Jet Fuels
(from oil-based feedstock such as soy beans)

3) Other More Challenging Fuels
(ethanol, methane, liquid hydrogen)

Alternative fuels include synthetic, bio-derived renewables, and more challenging fuels such as hydrogen.

level testing. This technology has been developed by WebCore Technologies Inc., under SBIR Phase I and II contracts with additional investment from the Aviation Safety Program.

Besides the immediate commercial transition of these two new advanced composites technologies for jet engine fan casings/containment system structures, both A&P Technology and WebCore Technologies are pursuing possible spinoff applications in other economic market sectors.

Alternative Fuels

NASA's aeronautics programs have a long history of pursuing technology directed at increasing the efficiency of modern aircraft and reducing emissions. In the 1970s and 1980s, when the cost of foreign oil dramatically increased and the country faced the possibility of fuel costing $2 per gallon (equivalent to about $5 per gallon in current year dollars), NASA began to aggressively fund programs to reduce fuel consumption. Research and

technology development programs focused on lightweight composite materials to enable much lighter airframes. Advanced propulsion programs were also created to improve the efficiency of existing gas turbine engines, and to explore new concepts such as an advanced, high-speed turbo propeller.

Today, the future of aviation is facing multiple challenges. Current projections show that air travel will increase by a factor of two to three sometime in the next decade, with a proportional increase in fuel demand. The latter will occur at a time when the world demand for oil will sharply increase, making U.S. dependence on foreign

oil a more serious issue than today. Similarly, if nothing is done, emissions from aircraft will also increase by a factor of two to three, including both nitrogen oxides (NOx) and carbon dioxide (CO_2). Today, aircraft only contribute about 3 percent of CO_2 emissions with no significant impact on the environment. Though the percentage may remain small in the future, the total amount of aircraft-emitted CO_2 will increase with the volume of air travel.

To address these issues, NASA's Fundamental Aeronautics Program conducts research into alternative fuels that are available from domestic resources (e.g., agricultural products and coal) and that produce less potentially harmful emissions. One approach is the Fischer-Tropsch process developed by German scientists in the 1920s. Today, it is still one of the more promising processes to produce fuels from non-petroleum raw materials. In the United States, the process has been used to produce ethanol from corn, as well as gaseous and liquid fuel from coal. However, there is no process, feedstock, or final product that best meets the aviation requirements for efficiency, low emissions, and cost.

This is one focus of NASA's Subsonic Fixed Wing Project. In addition to conducting in-house research, NASA will engage both industry and academia to research and develop better production processes and fuels that can be based on domestic feedstock. NASA also conducts research on advanced combustion processes and controls to optimize the benefit of alternative fuels. This includes defining the chemical kinetics and characterization of the alternative fuels, and running them in small labs before testing on engines.

If this work is successful, it has the potential to generate significant commercial opportunities that can extend well beyond the aviation community. NASA will not produce new fuels, nor will the Federal government, in general. The knowledge and capability to produce alternative fuels will reside in the private sector, including the engineering expertise and manufacturing infrastructure to support large scale production. Some of this economic spinoff will be realized by existing large-scale energy providers. However, as with any new technological development, new players also emerge—both small and large. Further, the basic technology for alternative aviation fuels is likely to apply to other markets as well. In fact, the broader the general markets for the technology and fuel products, the lower the costs and greater the benefits to the aviation community.

Space Operations Mission Directorate

The Space Operations Mission Directorate provides NASA with leadership and management of the Agency's space operations related to human exploration in and beyond low-Earth orbit. Space Operations also oversees low-level requirements development, policy, and programmatic oversight. Current exploration activities in low-Earth orbit include the space shuttle and International Space Station programs. The directorate is similarly responsible for Agency leadership and management of NASA space operations related to launch services, space transportation, and space communications in support of both human and robotic exploration programs. Its main challenges include: completing assembly of the ISS; utilizing, operating, and sustaining the ISS; commercial space launch acquisition; future space communications architecture; and transition from the space shuttle to future launch vehicles.

Opening a New Chapter in Space Exploration

Future NASA astronauts who land on the Moon will owe their success in part to the men and women of the U.S. Gulf Coast, who are already at work on the next generation of space travel. Stennis Space Center, in Mississippi, and NASA's Michoud Assembly Facility, in New Orleans, both will have critical roles in the Constellation Program, which aims to land astronauts on the Moon by the end of the next decade.

Stennis broke ground for a new rocket engine test stand that will provide altitude testing for the J-2X engine. The engine will power the upper stages of NASA's Ares I and Ares V rockets. NASA Deputy Administrator Shana Dale was joined by Mississippi Governor Haley Barbour, U.S. Senator Thad Cochran, U.S. Senator Trent Lott, and U.S. Representative Gene Taylor for the landmark occasion. Also participating were NASA Associate Administrator for Exploration Systems Scott Horowitz, and Stennis Center Director Richard Gilbrech, named to succeed Horowitz in October 2007. Pratt & Whitney Rocketdyne president Jim Maser took part as well.

This engineer's concept drawing of the A-3 Test Stand shows the 300-foot-tall structure's open steel frame and large exhaust diffuser.

"Groundbreakings are about new beginnings," said Dale. "The first stand was erected at Stennis to test the Saturn V rocket of the Apollo Program. Testing of the space shuttle engines began here in the mid 1970s. And today, we're breaking ground for a new test stand, for the new spacecraft of a new era of exploration." The Ares I and Ares V rockets are being developed as part of NASA's Constellation Program. Constellation spacecraft will be used to send astronauts to the International Space Station (ISS), return humans to the Moon, and eventually journey to Mars.

"This is our generation's turn, our time to go to the Moon," said Gilbrech. "One of the key steps is building the A-3 test stand. The J-2X engine has a unique set of test requirements. The best way to meet them is with the A-3."

The A-3 stand will be a 300-foot-tall, open steel frame structure located south of the existing A-1 test stand. Its 19-acre site in Stennis' A Complex will include a test control center, propellant barge docks, and access roadways. The test stand will allow engineers to simulate conditions at different altitudes by generating steam to reduce pressure in the test cell. Testing on the A-3 stand is scheduled to begin in late 2010.

"The engines will be assembled here at Stennis, then subjected to rigorous, expert testing," Dale said. "After that, those engines and the rockets they will power will travel to Cape Canaveral. Then the finished spacecraft will lift off, headed for a new destination and a new era of exploration."

NASA Report Details Education Concept for International Space Station National Laboratory

The 2005 NASA Authorization Act designated the U.S. segment of the ISS as a national laboratory and directed NASA to develop a plan to "increase the utilization of the ISS by other Federal entities and the private sector...." As the Nation's newest national laboratory, the ISS will further strengthen relationships among NASA, other Federal entities, and private sector leaders in the pursuit of national priorities for the advancement of science, technology, engineering, and mathematics. The ISS National Laboratory will also open new paths for the exploration and economic development of space.

- The ISS represents a unique and highly visible national asset with surplus capacity available for a wide spectrum of applications.

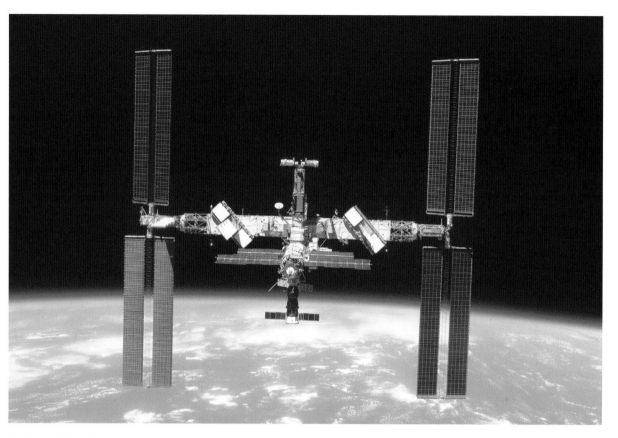

The International Space Station, against the blackness of space and Earth's horizon, at the end of STS-117's mission on June 19, 2007.

- The national laboratory concept is an opportunity to expand the U.S. economy in space-based research, applications, and operations.

- NASA will continue to cover the cost of operating and maintaining the ISS, and is highly motivated to work with other agencies and organizations to pursue applications.

A task force representing seven Federal agencies, including NASA, the National Science Foundation, and the Department of Education, has developed a strategy for using the ISS National Laboratory as a venue for further inspiring teachers and students in the areas of science, technology, engineering, and mathematics. The task force's education development concept looks at ways to use the space station's U.S. segment to support future projects and develop partnerships for education payloads, or experiments, with other Federal agencies. Some ideas include establishing an education working group with representatives from Federal agencies responsible for soliciting, selecting, and submitting education payloads; linking education activities with ongoing science investigations; and developing a payload rack filled with education-related materials and equipment.

For more than 6 years, students have successfully conducted classroom versions of station experiments and learned about the weightlessness of space through on-orbit demonstrations by crew members. The "International Space Station National Laboratory Education Concept Development Report" is the first phase in planning expanded educational use of the space station by multiple organizations as part of the designation of the ISS as a national laboratory.

Operation Dark Dune

On Launch Pad 39A at NASA's Kennedy Space Center, the Space Shuttle Endeavour sat bathed in glowing light, silhouetting the vehicle against the dark night sky over the seaside complex. An awesome scene in an idyllic location, and a striking counterpoint to the nesting sea turtles and their newly hatched babies on the nearby shore. During their summer nesting season, these turtles emerge from the ocean along the pristine beach within 200 yards of the space shuttle launch pads. The light emanating from the pads can deter the adults from coming ashore to lay their eggs and disorient the hatchlings as they emerge from their nests and head toward the moonlit sea. To help preserve the balance of its natural surroundings, Kennedy's environmental management system has as one of its goals to minimize controllable impacts to wildlife, including the nesting sea turtles.

While their height normally provides a necessary buffer between the launch pads and the shoreline, the dunes along Florida's Space Coast have been severely eroded in some spots by hurricanes, particularly during the 2004 season. That year, the Space Center was impacted by two hurricanes just 3 weeks apart, and while some dune restoration was completed, and more is planned, some stop-gap measures were needed until the nesting season ends at the beginning of November.

Enter some inventive individuals with a novel idea: Use what they have on hand to help block the launch pad lights, so the nesting process can continue undisturbed. As those charged with helping to protect the environmental balance debated how to shield the beach from the lights, Doug Scheidt, of Dynamac Corporation, Kennedy's life sciences support contractor, had this idea: Use freight train boxcars to shade the dunes. Admittedly a proverbial shot in the dark, freight train cars are about the right height to shade the dunes in the most severely eroded spots, and since the Space Center has a rail line that parallels the

Atlantis and a turtle race out to Pad 39B for mission STS-81.

beach, it was a viable solution that would also avoid the pitfalls inherent in trying to erect some type of temporary barriers that would require permits and funding.

Uniquely bringing together employees from both the operations and environmental sides of Kennedy's management team, the railcar idea took shape. The cars were big enough and mobile, and some scheduled to be removed from service were conveniently parked just a few miles away from the launch pads. The solution was ideal—quick, easy, and cheap—and the project that had been affectionately dubbed "Operation Dark Dune" strategically relocated 25 railcars to their temporary seaside location.

"As a former environmental protection specialist at Kennedy, I realize how fine a line it is between our operations and the protection of our natural resources," said Propellants Mobile Equipment Manager Gail Villanueva, who is in charge of the railcars. "I was happy I was in a position to help out, although the request was unique, to say the least."

The relationship between space exploration and nature goes back as far as the Space Program's roots in the region. When the Kennedy Space Center was carved out along the vast coastal area, its first director, Kurt H. Debus, arranged for a large portion of the Center to be designated as a wildlife refuge. Known as the Merritt Island Wildlife Refuge, it now encompasses 140,000 acres and is managed by the U.S. Department of the Interior's Fish and Wildlife Service. Kennedy also borders the Canaveral National Seashore, which provides an important nesting area for the sea turtles.

If innovative thinkers at the Space Center can continue to come up with creative solutions like Operation Dark Dune, the Center's dedication to the delicate balance between nature and space exploration will continue to flourish. ❖

Education News

NASA's missions to space are the result of scientific expertise and technical excellence, qualities that are dependent on sound educational backgrounds. NASA gives back to the educational community in order to groom the next generation of explorers.

Education News

NASA Sets Sights on the Next Generation of Explorers

NASA's Education Office has released a new framework for the academic community to prepare the next generation of explorers and innovators. The NASA Education Strategic Coordination Framework highlights Agency content, people, and facilities as the foundation for sponsored educational opportunities, while developing new non-traditional partnerships.

"NASA's education framework continues our investment in students, educators, and quality programs in order to engage and prepare our future scientists, engineers, and technicians. The exciting and compelling NASA missions truly inspire the next generation," said Dr. Joyce Winterton, assistant administrator for education.

The framework identifies three priorities for NASA to work with academia, industry, and informal educators to foster increased studies in science, technology, engineering, and mathematics. NASA's education priorities include strengthening the Nation's workforce, attracting and retaining students, and engaging America in NASA's missions.

Another key element of the new education framework is to involve partners and establish strategic alliances to work with NASA to inspire and engage the Nation's youth.

The Agency strives to reach students at every level in the educational system. NASA remains committed to engaging and retaining underrepresented and underserved communities of students, educators, and researchers in its education programs.

NASA Partners in American Indian Science and Engineering Education

NASA and the American Indian Higher Education Consortium (AIHEC) announced the launch of the NASA AIHEC Summer Research Experience Program. The program is a strategic approach to inspiring young American Indians to pursue careers in science and engineering.

Sixty participants representing 14 tribal colleges and universities took part in the program at 7 NASA centers this summer. Participants were assigned to research and engineering teams exploring robotics, 3-D design, geospatial data analysis, and astrobiology, while fostering long-term relationships with their research mentors at the field centers.

"American Indians are very underrepresented in the fields of science and engineering," said Dr. Gerald Gipp, executive director of AIHEC. "The program is a critical step toward changing that equation by encouraging young American Indians to pursue careers in science and engineering while also building a welcoming environment that nurtures their career path."

Tribal colleges with students and faculty participating in the program included: Blackfeet Community College, Browning, Montana; College of Menominee Nation, Keshena, Wisconsin; Crownpoint Institute of Technology, Crownpoint, New Mexico; Diné College, Tsaile, Arizona; Haskell Indian Nations University, Lawrence, Kansas; Keweenaw Bay Ojibwa Community College, Baraga, Michigan; Leech Lake Tribal College, Cass Lake, Minnesota; Little Priest Tribal College, Winnebago, Nebraska; Northwest Indian College, Bellingham, Washington; Oglala Lakota College, Kyle, South Dakota; Salish Kootenai College, Pablo, Montana; Southwestern Indian Polytechnic Institute, Albuquerque, New Mexico; Tohono O'odham Community College, Sells, Arizona; and United Tribes Technical College, Bismarck, North Dakota.

NASA and Honeywell Win Top Award for Science Education Initiative

NASA and Honeywell International Inc.'s joint science education effort "FMA Live!" was recognized as the top community outreach program in the United States during the recent Promotional Marketing Association's 2006 Reggie Awards.

The association awards annually identify and honor the best integrated U.S. marketing programs. The "FMA Live!" program received a "Gold Reggie" award in the "Cause/Community Outreach" category.

The program is part of a national partnership between NASA and Honeywell Hometown Solutions. The effort engages middle school students in the wonders of science, technology, and math through innovative programs and by highlighting the relevance of natural sciences encountered during their daily lives.

"FMA Live!" was named for Sir Isaac Newton's second law of motion (force = mass × acceleration). The program uses interactive science demonstrations, professional actors, original songs, and music videos to teach middle school students Newton's three laws of motion and the universal law of gravity.

This interactive program addresses critical curriculum objectives to help students understand the Newtonian concepts and to improve their performance in the sciences. Created in 2004, the program has traveled 23,000 miles, visiting 153 schools in 32 states, reaching more than 73,000 students. The program's Web site, www.fmalive.com, provides classroom lesson plans and other educational material for math and science studies.

During each performance, students, teachers, and administrators interact with three professional actors on stage in front of a live audience to experience Newton's laws firsthand. A large VELCRO wall is used to demonstrate inertia; go-carts driven across the stage illustrate action and reaction; and wrestling and a huge soccer ball show that force is determined by mass multiplied by acceleration. All three of Newton's laws are demonstrated when a futuristic hover chair collides with a gigantic cream pie.

NASA Kids' Club Web Site Is Entertaining and Educational

The NASA Kids' Club Web site features animated, colorful, entertaining, and educational activities for children in kindergarten through fourth grade.

NASA partnerships are working to teach kids about space exploration through interactive and entertaining Web sites.

Interactive games on the site teach children about exploring space, building and launching rockets, keeping airplanes on schedule, and how comets travel through the solar system. The site is located at www.nasa.gov/kidsclub.

The site serves a dual purpose. Children can play games at home for entertainment, and educators can use it as a fun way to reach students in the classroom, the library, during after-school programs, or anywhere children and computers are together.

NASA's Educational Technology Services team at the Marshall Space Flight Center, in Huntsville, Alabama, developed and maintains Kids' Club. The site was designed in accordance with the 2004 National Education Technology Plan, "Toward a New Golden Age in American Education." Through the interactive site, content is aligned with educational standards that are customized to students' individual needs and interests.

NASA and AOL Team Up to Give Kids a Virtual Ride into Space

NASA and AOL have joined forces to bring the excitement and adventure of space exploration to young people. Through a Space Act Agreement, NASA's Office of Education and the Space Operations Mission Directorate will collaborate with AOL to create a series of live Web casts for its "KOL Expeditions: NASA Earth Crew Mission" site.

To kick off this new partnership, AOL's kid's online service, KOL, presented a Web cast during National Space Day at NASA's Goddard Space Flight Center.

During the show, students had an "out of this world" opportunity to talk with the Expedition 13 crew members aboard the International Space Station. Expedition 13 commander Pavel Vinogradov and flight engineer Jeff Williams began their 6-month mission on the station March 31, 2006. Additional Web cast participants included NASA astronauts Paul Richards and Ricky Arnold, and Goddard's chief scientist, James Garvin.

"We're thrilled to partner with NASA as part of our KOL Expeditions program, enabling millions of kids to discover new frontiers by interacting with NASA scientists and astronauts," said Mark Stevens, KOL education director and general manager of AOL@SCHOOL.

KOL Expeditions is a new site designed to provide kids, parents, and teachers with a platform to make learning fun through interactive missions, video content, and other special activities. The NASA/AOL station Web cast, the second initiative on the site, lays the groundwork for what will be a series about space exploration.

The site also features educational materials and interactive activities to inspire elementary-aged kids to explore the world of science and space.

National Space Day, sponsored by Lockheed Martin Corporation, started in 1997 as a grassroots educational initiative to be held each year in early May. The goal of Space Day is to use the excitement of space exploration to encourage students' interest in science, technology, engineering, and mathematics.

Hundreds of thousands of teachers and millions of students have participated in Space Day events in 21 countries. The program, which is supported by more than 70 official partner institutions, has been honored with the Space Foundation's Education Achievement Award.

NASA Launches New Education Initiative for Minority Institutions

NASA kicked off a new initiative with the United Negro College Fund Special Programs Corporation this year. The initiative will give researchers and students from minority institutions direct access to NASA facilities, scientists, and capabilities.

Funded by a $3.5 million grant from NASA, the corporation will establish the NASA Science and Technology Institute for Minority Institutions. The institute will be in the NASA Research Park at Ames Research Center.

Focused on science, technology, engineering, and mathematics, the institute will bring together the talent and expertise of historically black colleges and universities, Hispanic-serving institutions, tribal colleges and universities, and other minority institutions through research-based fellowships, internships, co-ops, and grants.

"I am truly delighted that NASA is partnering with the corporation to establish this revolutionary new institute," said S. Pete Worden, director of Ames. "This joint venture will give minority students and researchers access to NASA and the opportunity to collaborate with researchers in the surrounding community of universities, high-tech research, and development companies."

The goal of the initiative is to provide professional development that will prepare faculty, students, researchers, visionaries, and entrepreneurs to become highly skilled science and technology leaders and managers, and to compete in the national and global workforce.

"The establishment of this institute truly demonstrates and highlights NASA's continuing commitment to promoting science, technology, engineering, and mathematics excellence in the minority higher education community," said Aaron R. Andrews, president and CEO of the corporation.

A Stellar Educator...

Educator Astronaut and Mission Specialist Barbara Morgan made her first spaceflight in August 2007 on STS-118. Educator Astronauts are K-12 teachers who are selected by NASA to become fully qualified astronauts, and will help NASA develop new ways to connect space exploration with the classroom and inspire the next generation of explorers. They also have the task of using their out-of-this-world experiences to help other teachers excite students about science, technology, engineering, and mathematics.

Barbara Morgan began her 26-year teaching career in 1974 on the Flathead Indian Reservation at Arlee Elementary School in Arlee, Montana, where she taught remedial reading and math. She was selected as the backup candidate for the NASA Teacher in Space Program on July 19, 1985, training at Johnson Space Center with Christa McAuliffe and the Challenger crew.

Following the Challenger accident, Morgan assumed the duties of Teacher in Space Designee, including working with NASA as a speaker visiting educational organizations throughout the country, educational consulting and curriculum design, and serving on the National Science Foundation's Federal Task Force for Women and Minorities in Science and Engineering. In the fall of 1986, she resumed her teaching career, teaching second and third grades at McCall-Donnelly Elementary School, in McCall, Idaho, while continuing to work with NASA's Education Division in the Office of Human Resources and Education.

In 1988, students at McCall-Donnelly Elementary took part in the NASA Orbiter-Naming Project, suggesting a name for NASA's newest orbiter. Little did they know, they were helping to name the very shuttle that would one day carry one of their teachers into space. McCall-Donnelly was one of more than 6,000 U.S. schools that participated in the competition, and the school's submission, "Endeavour," was the most popular entry.

Morgan was selected by NASA as a mission specialist in 1998, and again reported to Johnson. Following the completion of 2 years of training and evaluation, she served in the International Space Station Operations, Capsule Communicator (CAPCOM), and Robotics branches of the Astronaut Office. She was assigned to the crew of STS-118 as a mission specialist, one of the robotics operators responsible for controlling the shuttle's Canadian-built robotic arm and the station's robotic arm during spacewalks and other activities.

Endeavour's first flight in more than 4 years, STS-118 worked with the Expedition 15 crew to continue construction of the International Space Station with the addition of the Starboard 5 truss. The crew conducted four spacewalks, which included the truss installation

Barbara Morgan became the first teacher in space when she flew on the STS-118 mission. The crew members of STS-118 included (from the left) mission specialists Richard A. (Rick) Mastracchio, Barbara R. Morgan, pilot Charles O. Hobaugh, commander Scott J. Kelly, and mission specialists Tracy E. Caldwell, Canadian Space Agency's Dafydd R. (Dave) Williams, and Alvin Drew, Jr.

Astronaut Barbara R. Morgan, STS-118 mission specialist, smiles for the camera while working on the middeck of Space Shuttle Endeavour during Flight Day Two activities.

and replacement of an attitude control gyroscope, before safely returning on August 21, 2007. STS-118 was the 119th space shuttle flight and the 22nd shuttle mission to visit the station.

When asked by a student during her time in space which of her occupations she preferred, Morgan's response encapsulated the enthusiasm and idealism that has marked her time both as an educator and astronaut, saying, "Do I have to choose one, or can I do both, please? Actually, both are excellent jobs, and they're both very, very similar. Both you're exploring, you're learning, you're discovering, and you're sharing. And the only difference really to me is that as an astronaut you do that in space, and as a

teacher you get to do that with students. And they're both wonderful jobs. I highly recommend both."

...and a Stellar Student

The 45th Robert H. Goddard Memorial Symposium held March 20-21, 2007, drew attention to significant milestones, events, missions and projects underway at NASA, and the event highlighted potential future NASA leaders in the student poster sessions. Students used the sessions to share their research and contributions as members of the NASA team, and while each student displayed exemplary knowledge and potential, one high-school senior's inspired efforts exemplified the type of accom-

plishments that can propel a nation to explore the Moon, Mars, and beyond.

Anna Cyganowski showed a driven, proactive nature certain to propel her to great heights of achievement. In her application to the High School Internship Project (HIP) sponsored by Goddard Space Flight Center, Cyganowski indicated an interest both in robotics and the finite research area of metal whiskers and dendrites—pesky metal filaments that grow on circuit boards and are often the culprit of expensive electrical shorts in satellites, pacemakers, and other critical electronic equipment.

As the HIP duration is only 7 weeks, students traditionally receive a single assignment. Cyganowski was set to support the NASA Robotics Academy, but she, in her pursuit to obtain as valuable an experience as possible, wanted more. She sought out and was granted permission to pursue a second internship assignment in metal whiskers and dendrites, and after coordination with HIP's manager, her mentor in the NASA Robotics Academy, and the world-renowned researchers, Dr. Henning Leidecker and Jay Brusse, her ideal internship was set.

Cyganowski continued her investigation of metal whiskers and dendrites well past the end of her inspirational and beneficial experience with the NASA Robotics Academy. Her passion for science and engineering has stimulated interest in technology among other students at her high school, stirred interest in the local community and among professional scientists, and garnered notable accolades:

- At the 2007 Goddard Symposium Poster Session, Cyganowski was one of two high school students selected to present. As part of her research, she documented findings through photographs and videos, which the Parts, Packaging, and Assembly Technologies Office at Goddard now uses to illustrate the difference between metal whiskers and metal dendrites.

- Among 1,705 entrants to the Intel Science Talent Search 2007 competition—which considers itself "America's oldest and most prestigious high school science competition" and the "Junior Nobel Prize"—Cyganowski was named one of three hundred semifinalists for her project, "Nickel Cadmium Batteries: A Medium for the Study of Metal Whiskers and Dendrites." The achievement earned her an award of $1,000 and an additional $1,000 for her high school, Notre Dame Preparatory School, in Towson, Maryland, to fund future programs for exemplary students.

- The 2007 National Space Club Scholars Program awarded Cyganowski the coveted Olin E. Teague Memorial Scholarship for 2007.

- Baltimore County executive Jim Smith declared Feb. 5, 2007, "Anna Cyganowski Day."

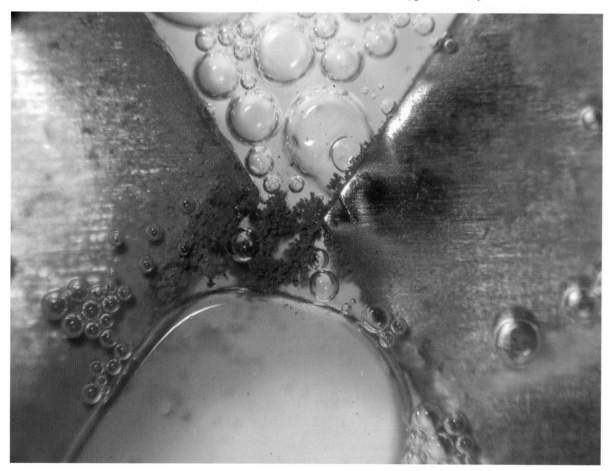

The formation of copper dendrites, from the report "Nickel Cadmium Batteries: A Medium for the Study of Metal Whiskers and Dendrites" by Anna Cyganowski, Jay Brusse, and Dr. Henning Leidecker.

While busy with numerous other activities, Cyganowski maintains a focus on her academics and future career, and her accomplishments have already contributed to NASA science and the Agency's future.

NASA's Newest Explorers

NASA welcomed 25 new NASA Explorer School teams on May 11, 2007, in a partnership to inspire students in science, technology, engineering, and mathematics. A "pipeline" strategic initiative, Explorer Schools promote and support the incorporation of NASA content and programs into science, technology, and mathematics curricula in classroom grades four through nine across the United States. Targeting underserved populations in diverse geographic locations, NASA Explorer Schools bring together educators, administrators, students, and families in sustained involvement with NASA's education programs and provide unique opportunities designed to engage and educate the future scientists who may someday help advance U.S. scientific interests through space exploration.

To compete effectively for the minds, imaginations, and career ambitions of America's young people, NASA focuses on engaging and retaining students in educational efforts that encourage the pursuit of disciplines critical to NASA's future missions. Students engage in research, problem solving, and design challenges relating to NASA's missions, which involves them in the study of science, technology, engineering, and mathematics. Activities encourage the use of scientific tools and methods, and challenges are grade-specific, supporting national and state standards. In addition, in-flight opportunities and competitions provide access to unique NASA resources and personnel.

To begin the formal partnership, a team composed of full-time teachers and a school administrator attends a 1-week professional development workshop at their respective NASA center that provides opportunities to begin integration of NASA content into existing school curricula. Teams develop and implement a 3-year action plan to address local challenges in science, technology, and mathematics education, and representatives from NASA centers help kick off the program with presentations at the schools in their region during the coming school year.

Currently, 200 teams are involved in the project, representing all 50 states, the District of Columbia, Puerto Rico, and the U.S. Virgin Islands. A grant to each school supports the purchase of technology tools, online services, and in-service support for the integration of technology applications to engage students in science and mathematics investigations. Selected schools are eligible to receive up to $17,500 during the 3-year partnership, and the project also provides educators and students with content-specific activities that can be used in many local and state curricula to excite students about science and math. The NASA Explorer Schools includes NASA resources; science, technology, and mathematics investigations; collaborative tools; and opportunities to share student and school program results. ❖

VELCRO® is a registered trademark of Velcro Industries B.V.

Professional development workshops for teams of teachers and school administrators begin the integration of NASA content into existing school curricula.

Student programs provide opportunities for active participation in research, problem solving, and design challenges relating to NASA's missions.

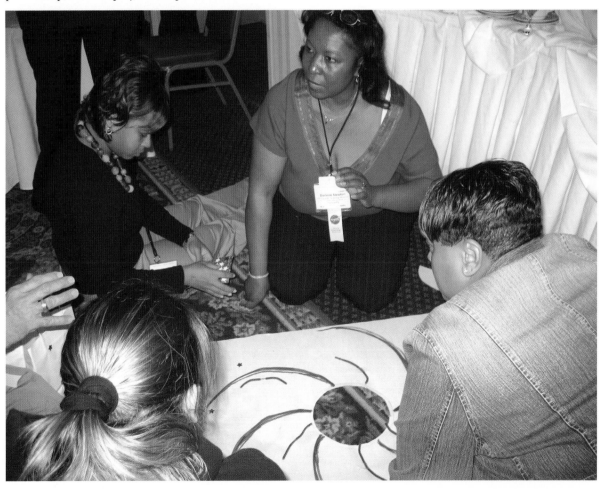

Partnership News

NASA cultivates non-commercial relationships with private industry, academia, and other government agencies to bring its science back down to Earth. By contributing time, facilities, and a wealth of technical expertise, NASA enriches the lives of people everywhere with these successful partnerships.

Partnership News

The Innovative Partnerships Program aims to provide leveraged technology for NASA's mission directorates, programs, and projects through investments and technology partnerships with industry, academia, government agencies, and national laboratories. The following stories highlight some of the exceptional results of these many partnerships.

NASA Receives Award for Excellence in Technology Transfer

NASA was honored March 1, 2007, for successfully conducting a broad range of technology transfer activities. The International Marketplace and Conference for Business Development through Technology Transfer (IPTEC), of St. Albans, England, presented the award to NASA during its conference in Cannes, France. Doug Comstock, NASA's director of the Innovative Partnerships Program, accepted the award on behalf of the Agency.

"When we collectively engage in space exploration, we invest not only in the successful navigation of the unknown but also the innovations that improve our very quality of daily life," said Shana Dale, NASA's deputy administrator. "We congratulate our program's accomplishments of contributing to the high-quality technology transfers that benefit exploration while complimenting American industry's ability to provide benefits for our entire society."

During its annual conference, IPTEC presents three awards, one each to the public, private, and academic sectors. IPTEC's advisory board, comprised of representatives from corporations such as the General Electric Company, Microsoft Corporation, and Ericsson Inc., recommended the recipients of the awards. At the IPTEC conference, many of the world's leading experts in technology transfer discuss the latest corporate and government technology transfer strategies and learn about successful licensing programs and practices.

"This is an important recognition for NASA, because we take seriously the transfer of technology from our unique space and aeronautics missions into productive societal use," said Comstock.

NASA and Universities Join to Fight Diabetes

A NASA image processing technology used to explore orbital images of Earth and distant worlds is being modified for diabetes research.

Scientists at The George Washington University and Cornell University helped modify the technology, which has greatly increased the speed of the research. "NASA technology combined with our modifications has provided

Photomicrograph of a sliced rat beta cell that has been processed with the modified NASA imaging technology. Insulin granules are the dark black spots surrounded by a white area called a halo. Large cells have hundreds of insulin granules. The colored borders around the granules are labels added to identify them and classify how they appear.

us with new tools for fighting diabetes," said Murray Loew, director of the Biomedical Engineering program and professor of engineering at The George Washington University's School of Engineering and Applied Science.

Diabetes afflicts more than 20 million Americans. It is caused by the body's inability to regulate glucose, a sugar that cells use for energy. The hormone insulin regulates blood glucose levels by unlocking the interior of cells and allowing glucose in the blood to pass through the cell wall. Insulin is manufactured in beta cells in the pancreas. Microscopic structures called granules carry insulin toward the wall of the beta cells, where it is secreted in response to glucose levels in the blood.

Two types of diabetes exist. In Type I diabetes, pancreatic cells are destroyed. In Type II diabetes, either pancreatic cells do not secrete enough insulin, or cells in the body lose their responsiveness to insulin, or both problems happen at once. Both types of diabetes cause glucose to build up in the blood instead of being delivered to the interior of cells, where it is needed or would be stored. Potential effects include coma, heart disease, kidney damage, nerve damage, blindness, and loss of limbs.

In the research, the team analyzed electron photomicrographs (images from an electron microscope) of beta cells from rats.

The original NASA technology helps scientists to classify picture elements (pixels) and identify different types of landforms, geology, and vegetation. In the laboratory, it has been adapted to identify biological structures, the insulin granules, in electron photomicrographs. The research team observed the number, size, and position of insulin granules in the beta cells in response to glucose.

"Previously, the analysis of each electron micrograph took an assistant several hours to complete. Now, with the image processing software, we can automatically analyze several dozen electron micrographs overnight," said Tim McClanahan, a scientist at NASA's Goddard Space Flight Center.

"We plan on an extensive collaboration in the future. The potential for this research is excellent," said Geoffrey Sharp, a diabetes expert in the Department of Molecular Medicine at Cornell University. The team has submitted proposals to the National Institutes of Health (NIH) and the American Diabetes Association to further validate the technology with additional data and to extend the work to identify and characterize other microscopic cellular structures.

The research is being funded by Goddard's part-time graduate study program, NIH, and the Juvenile Diabetes Research Foundation International.

NASA Explains Puzzling Impact of Polluted Skies on Climate

NASA scientists have determined that the formation of clouds is affected by the lightness or darkness of air pollution particles. This also impacts Earth's climate.

In a breakthrough study published in the online edition of Science, scientists explain why aerosols—tiny particles suspended in air pollution and smoke—sometimes stop clouds from forming and, in other cases, increase cloud cover. Clouds not only deliver water around the globe, they also help regulate how much of the Sun's warmth the planet holds. The capacity of air pollution to absorb energy from the Sun is the key.

"When the overall mixture of aerosol particles in pollution absorbs more sunlight, it is more effective at preventing clouds from forming. When pollutant aerosols are lighter in color and absorb less energy, they have the opposite effect and actually help clouds to form," said Goddard's Lorraine Remer. Remer worked closely with the study's lead author, the late Yoram Kaufman, also of Goddard, on previous research into this perplexing "aerosol effect."

The effect of the planet's constantly changing cloud cover has long been a problem for climate scientists. How clouds change in response to greenhouse-gas warming and air pollution will have a major impact on future climate.

Clouds help regulate the Earth's climate by reflecting sunlight into space, thus cooling the surface. When cloud patterns change, they modify the Earth's energy balance in turn, and temperatures on the Earth's surface.

Using this new understanding of how aerosol pollution influences cloud cover, Kaufman and co-author Ilan Koren, of Israel's Weizmann Institute of Science, estimate the impact worldwide could be as much as a 5-percent net increase in cloud cover. In polluted areas, these cloud changes can change the availability of fresh water and regional temperatures.

In previous research by the authors, the opposite effects that aerosols have on clouds were seen in different parts of the world using data from NASA satellites. These observations alone, however, could not confirm that the aerosols themselves were causing the clouds to change.

To tackle this problem, Kaufman and Koren assembled a massive database of global observations that strongly suggests it is the darkness (absorbs sunlight) or brightness (reflects sunlight) of aerosol pollution that causes pollution to act as a cloud killer or a cloud maker. These mesasurements were culled from the NASA-sponsored Aerosol Robotic Network (AERONET) of ground-based instruments at nearly 200 sites worldwide.

No matter where in the world the measurements were taken or in what season, scientists saw the same pattern. There were lots of clouds when light-reflecting pollution filled the air, but many fewer clouds were recorded in the presence of light-absorbing aerosols.

NASA's satellites, computer models, and technology will continue to advance the understanding of how aerosol pollution affects the Earth's climate. NASA's "A-Train" of formation-flying satellites, now with the cloud-piercing instruments onboard the CloudSat and CALIPSO spacecrafts, is helping to answer challenging questions such as the role of clouds in global warming and the influence of aerosols on rainfall and hurricanes.

NASA Assists Search for Ivory-Billed Woodpecker

Unlike its more famous cartoon cousin, Woody Woodpecker, the ivory-billed woodpecker is thought to be extinct, or so most experts have believed for over half a century.

Recently, though, scientists from NASA and the University of Maryland launched a project to identify possible areas where the woodpecker, one of the largest and most regal species, might be living. Finding these habitat areas will guide future searches for the bird and help determine if it is really extinct or has maintained an elusive existence.

The question of whether the species still exists started when a kayaker reported spotting the woodpecker along

Artist's rendering of the ivory-billed woodpecker. If the bird does exist, NASA's Laser Vegetation Imaging Sensor could help find it.

Arkansas' Cache River in 2004. That sighting spawned an intensive search for the species by wildlife conservationists, bird watchers, field biologists, and others.

In June 2006, a research aircraft flew over delta regions of the lower Mississippi River to track possible areas of habitat suitable for the ivory-billed woodpecker, a project supported by the U.S. Fish and Wildlife Service and the U.S. Geological Survey.

Scientists from Goddard and the University of Maryland used NASA's Laser Vegetation Imaging Sensor (LVIS) onboard the aircraft. The instrument uses lasers that send pulses of energy to the Earth's surface. Photons of light from the lasers bounce off leaves, branches, and the ground and reflect back to the instrument. By analyzing these returned signals, scientists receive a direct measurement of the height of the forest's leaf-covered tree tops, the ground level below, and everything in between.

"LVIS is aiding this search effort far beyond what aircraft photos or satellite images can provide in the way of just a two-dimensional rendering of what's below," said Woody Turner, program scientist at NASA Headquarters. "The laser technology gives us the third dimension, enabling us to better assess the complex vegetation structure the plane flies over." The flights are the latest step in an effort spanning over 2 years to find absolute evidence that a bird once thought extinct continues to survive.

"We're trying to understand the environment where these birds live or used to live, using LVIS-plotted features like thickness of the ground vegetation and tree-leaf density, in combination with other factors like closeness to water and age of the forest, to determine where we might find them," said Turner.

"Through numerous studies, we have shown the effectiveness of the data generated by this sensor for many scientific uses, including carbon removal, fire prediction, and habitat identification," said LVIS project researcher Ralph Dubayah, a professor in the University of Maryland's Department of Geography. "Lidar technology like LVIS measures the vertical structure of the trees and ground, setting it apart from other remote-sensing systems that provide detailed horizontal information that tells us little about whether a green patch of forest is short or tall, for example. When identifying habitats, the vertical structure of the vegetation is of paramount importance to many species, including a bird like the ivory-bill."

The reported sighting of the ivory-billed woodpecker inspired a year-long search by more than 50 experts working together as part of the Big Woods Conservation Partnership, led by Cornell University's Cornell Laboratory of Ornithology and the Nature Conservancy. Researchers have followed reported sightings across a huge swath of the Southeastern United States, including the Gulf Coast, Alabama, and Florida.

In April 2005, that team published a report in the journal Science that at least one male ivory-bill still survived. However, some scientists have challenged whether it really was the ivory-billed woodpecker that was spotted. The NASA-University of Maryland project is designed to provide detailed habitat information that search teams will use for expanded efforts to find new evidence about the possible survival of the bird.

The project also has a broader application, according to Goddard's Bryan Blair, principal investigator for the project. "This field campaign is part of an effort to develop approaches that bring together many types of remote-sensing data for monitoring wildlife habitat."

The research team previously used NASA's LVIS to study wildlife habitats in old-growth forests in the Western United States and in rain forests abroad.

NASA Study Solves Ocean Plant Mystery

A NASA-sponsored study shows that by using a new technique, scientists can determine what limits the growth of ocean algae, or phytoplankton, and how this affects Earth's climate.

Phytoplankton is a microscopic ocean plant and an important part of the ocean food chain. By knowing what limits its growth, scientists can better understand how ecosystems respond to climate change.

The study focused on phytoplankton in the tropical Pacific Ocean. It is an area of the ocean that plays a particularly important role in regulating atmospheric carbon dioxide and the world's climate, in that it is the largest natural source of carbon dioxide to the atmosphere.

"We concluded that nitrogen is the primary element missing for algae growth and photosynthesis in the northern portion of the tropical Pacific, while it

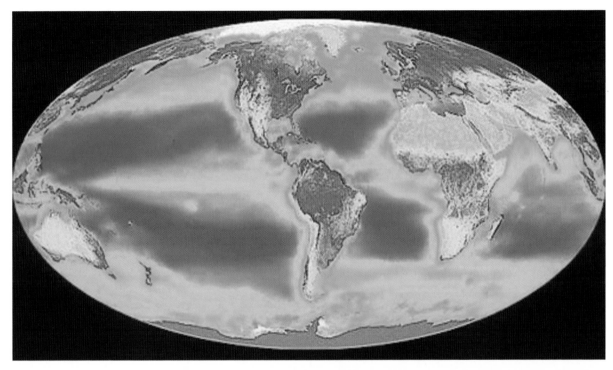

This image depicts amounts of plant life on Earth. On land, the dark greens show where there is abundant vegetation, and the tan colors show relatively sparse plant cover. In the ocean, red, yellow, and green areas show higher levels of phytoplankton, and these are regions of the ocean that are the most productive over time, while blue and purple areas show where there is very little phytoplankton.

will become more knowledgeable about where carbon is going and the impact of recreational, industrial, or commercial processes that use or produce carbon. Researchers better understand the Earth as an ecosystem and can incorporate these findings in future modeling, analysis, and predictions.

While satellite data from NASA's Sea-viewing Wide Field-of-view Sensor (SeaWiFS) played an important part in the study, the real cornerstone of the discovery was ship-based measurements of fluorescence.

Fluorescence occurs when plants absorb sunlight and some of that energy is given back off again as red light. Scientists looked at approximately 140,000 measurements

Satellite picture of phytoplankton bloom off Grand Banks, southeast of Newfoundland, using the OrbView-2/SeaWiFS.

was iron that was most lacking everywhere else," said Michael J. Behrenfeld, an ocean plant ecologist from Oregon State University.

Scientists determined that when phytoplankton is stressed from lack of iron, it appears greener, or healthier, than it really is. Normally, greener plants are growing faster than less green plants. When iron is lacking, enhanced greenness does not mean phytoplankton is growing better; it is actually under stress and unhealthy.

"Because we didn't know about this effect of iron stress on the greenness of algae or phytoplankton before, we have always assumed that equally green waters were equally productive," Behrenfeld said. "We now know this is not the case and that we have to treat areas lacking iron differently."

For the tropical Pacific, correction for this "iron-effect" decreases estimates of how much carbon ocean plants photosynthesize for the region by roughly 2 billion tons. This figure represents a tremendous amount of carbon that remains in the atmosphere that scientists previously thought was being removed.

The results of this study allow scientists using computer models to recreate the movement of carbon around the world much more accurately. Resource managers of fluorescence made from 1994 to 2006 along 36,040 miles of ship tracks. They found that phytoplankton gives off much more fluorescence when the plants do not have sufficient iron. It is this signal they used to fingerprint what parts of the ocean are iron-stressed and what parts are nitrogen-stressed.

It is important that scientists understand how ocean plants behave, because all plants play a critical role in maintaining a healthy planet. Plants annually absorb billions of tons of carbon dioxide from the atmosphere through photosynthesis and use this carbon to create the food that nearly all other organisms on Earth depend on for life.

Nutrients that make ocean plants thrive, such as nitrogen and phosphates, mostly come from the deep parts of the ocean, when water is mixed by the wind. Iron also can come from dust blowing in the air.

Approximately half of the photosynthesis on Earth occurs in the oceans, and ocean and land plants share the same basic requirements for photosynthesis and growth. These requirements include water, light, and nutrients. When these three are abundant, plants are abundant; when any one of them is missing, plants suffer.

NASA and Google to Bring Space Exploration Down to Earth

NASA's Ames Research Center and Google Inc. have signed a Space Act Agreement that formally establishes a relationship to work together on a variety of challenging technical problems ranging from large-scale data management and massively distributed computing, to human-computer interfaces.

As the first in a series of joint collaborations, Google and Ames will focus on making NASA's most useful information available on the Internet. Real-time weather visualization and forecasting, high-resolution 3-D maps of the Moon and Mars, as well as real-time tracking of the International Space Station and the space shuttle will be explored in the future.

"This agreement between NASA and Google will soon allow every American to experience a virtual flight over the surface of the Moon or through the canyons of Mars," said Michael Griffin, NASA administrator. "This innovative combination of information technology and space science will make NASA's space exploration work accessible to everyone," added Griffin.

"Partnering with NASA made perfect sense for Google, as it has a wealth of technical expertise and data that will be of great use to Google as we look to tackle many computing issues on behalf of our users," said Eric Schmidt, chief executive officer of Google. "We're pleased to move forward to collaborate on a variety of technical challenges through the signing of the Space Act Agreement."

Recently, teams from NASA and Google met to discuss the many challenging computer science problems facing both organizations and possible joint efforts that could help address them.

NASA and Google intend to collaborate in a variety of areas, including incorporating Agency data sets in the Google Earth mapping service, focusing on user studies and cognitive modeling for human-computer interaction, and on science data searching utilizing a variety of Google features and products.

"Our collaboration with Google will demonstrate that the private and public sectors can accomplish great things together," said S. Pete Worden, Ames center director. "I want NASA Ames to establish partnerships with the private sector that will encourage innovation, while advancing the Vision for Space Exploration and commercial interests," Worden added.

"NASA has collected and processed more information about our planet and universe than any other entity in

The partnership between NASA and Google Inc. will provide the public with access to much of NASA's information, including weather visualization and forecasting, high-resolution 3-D maps of the Moon and Mars, as well as real-time tracking of the International Space Station and the space shuttle.

the history of humanity," said Chris C. Kemp, director of strategic business development at Ames. "Even though this information was collected for the benefit of everyone, and much is in the public domain, the vast majority of this information is scattered and difficult for non-experts to access and to understand.

"We've worked hard over the past year to implement an agreement that enables NASA and Google to work closely together on a wide range of innovative collaborations," said Kemp. "We are bringing together some of the best research scientists and engineers to form teams to make more of NASA's vast information accessible."

NASA and Google also are finalizing details for additional collaborations that include joint research, products, facilities, education, and missions.

Spunky Satellite Yields Nobel Prize for NASA Scientist

In the early 1970s, a young NASA scientist had a crazy idea to build a strange-looking microwave satellite to test the Big Bang theory. After much stress and many false starts, his satellite finally launched in 1989 and by 1990 found nearly irrefutable evidence to support the Big Bang theory.

On October 3, 2006, the Nobel Prize Committee announced that this scientist, John Mather, of Goddard, would receive the 2006 Nobel Prize in Physics. He shares the prize with long-time colleague, George Smoot, of the U.S. Department of Energy's Lawrence Berkeley National Laboratory, in Berkeley, California.

Until fairly recently, very little was known about the origin of the universe. One theory, called the Big Bang, stated in the simplest of terms that, long ago, something happened, and about a billion years later stars and galaxies appeared. John Mather helped fill in the pieces. The satellite mission he led was the Cosmic Background Explorer (COBE).

As early as 1974, Mather was determined to build a satellite that could find evidence for the Big Bang and how stars and galaxies formed. The Big Bang theory grew out of Einstein's theory of general relativity and was developed by a Jesuit priest named Georges Lemaître and others in the 1920s.

The first striking evidence for the Big Bang came between 1963 and 1965, when Arno Penzias and Robert Wilson, of Bell Laboratories, stumbled upon some annoying microwave static interfering with their radio experiment. That interference, responsible for a sizable amount of static seen on your television set, turned out to be remnant radiation from the birth of the universe 13.8 billion years ago. Penzias and Wilson won the 1978 Nobel Prize in Physics for this discovery.

Mather and Smoot greatly advanced the field by precisely measuring the temperature and spectrum of this cosmic microwave background, the afterglow of the Big Bang that has cooled considerably but still lingers with us today. If our eyes could detect microwaves, we would see the entire sky bathed in this light.

The temperature they measured was 2.375 +/- 0.06 degrees Kelvin, or about minus 455 degrees Fahrenheit. More important, Mather and Smoot found slight temperature fluctuations within this near uniform light, which physicist Stephen Hawking, independent of the COBE team, called "the most important discovery of the century, if not of all time."

Why the hyperbole? The temperature variations (about 10 parts per million) make life possible. Without them, no stars or galaxies or planets would have formed. These variations—a little more heat here, a little less there—pointed to density differences, regions with a little more matter and a little less matter. Through gravity, over the course of billions of years—in a cosmic take on the "rich get richer" idea—those denser and warmer pockets attracted more matter and heat, which ultimately gave rise to the stars, galaxies, and hierarchical structure we see today.

The simplest model of the Big Bang cannot explain why stars formed, but the tweaked model for which Mather and Smoot found evidence can.

When Mather presented a chart of the first 9 minutes of COBE data at the 1990 meeting of the American Astronomical Society in Washington, D.C., he received

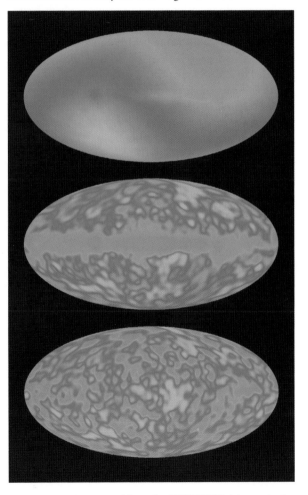

The images were created from the COBE DMR data products. Each image has been histogram equalized, giving a non-linear relation between color value and temperature.

a standing ovation. Scientists saw instantly how well the COBE data matched the temperature map predicted by theory. Rarely in science is a match between observation and theory so precise. The moment still gives Mather goose bumps today, he said.

Unlike the Hubble Space Telescope, COBE did not make optical images of stars and planets, which readily capture the public's imagination. As such, COBE never became a household name before its mission ended in 1994, yet its legacy is the Nobel Prize, the first to be awarded to a NASA scientist.

COBE carried three instruments. The first, the Far Infrared Absolute Spectrophotometer (FIRAS), measured the temperature and spectrum of the cosmic microwave background. Mather, the COBE mission project scientist, was the FIRAS principal investigator, and Richard Shafer, also of Goddard, was the deputy principal investigator.

The second instrument, the Differential Microwave Radiometer (DMR), measured the temperature variations, called anisotropy. George Smoot was the DMR principal investigator, and Charles L. Bennett, then at Goddard and now at Johns Hopkins University, was the deputy principal investigator.

The third instrument, the Diffuse Infrared Background Experiment (DIRBE), measured the cosmic infrared background, the "core sample" of the universe, containing the cumulative emissions of stars and galaxies dating back to the epoch of first light hundreds of millions of years after the Big Bang. The result was surprising: The universe has produced twice as much light as had been thought, and hidden it from view. A previously unknown population of galaxies made this light. Michael Hauser, then at Goddard and now at the Space Telescope Science Institute, was the DIRBE principal investigator. Tom Kelsall, of Goddard, was the deputy principal investigator.

NASA's Wilkinson Microwave Anisotropy Probe, now in orbit, builds on the COBE legacy, exploring in far greater detail the temperature variations the COBE discovered—quite possibly the stuff of future Nobel Prizes.

In August 2006, Mather and the COBE team won the 2006 Gruber Cosmology Prize, also for the Big Bang discoveries. Along with the scientists mentioned above, the recipients of this award include the members of the COBE Science Working Group: Eli Dwek, S. Harvey Moseley, Robert F. Silverberg, and Nancy Boggess (retired), of Goddard; Edward Cheng, formerly with Goddard and now president of Conceptual Analytics LLC; Samuel Gulkis and Michael A. Janssen, of NASA's Jet Propulsion Laboratory (JPL); Rainer Weiss, of the Massachusetts Institute of Technology; Stephan Meyer, of the University of Chicago; Philip Lubin, of the University of California, Santa Barbara; Edward Wright, of the University of California, Los Angeles; Thomas Murdock, of Frontier Technology Inc.; and the estate of the late David T. Wilkinson, of Princeton University.

NASA Probes the Sources of the World's Tiny Pollutants

Pinpointing pollutant sources is an important part of the ongoing battle to improve air quality and to understand its impact on climate. Scientists using NASA data recently tracked the path and distribution of aerosols—tiny particles suspended in the air—to link their region of origin and source type with their tendencies to warm or cool the atmosphere.

By altering the amount of solar energy that reaches the Earth's surface, aerosols influence both regional and global climate, but their impact is difficult to measure, because most only stay airborne for about a week, while greenhouse gasses can persist in the atmosphere for decades. In a new study, researchers investigated the sources of aerosols and how different types of aerosols influence climate.

"This study offers details on the aerosol source regions and emission source types that policy makers could target to most effectively combat climate change," said Dorothy Koch, lead author and an atmospheric scientist at Columbia University and NASA's Goddard Institute for Space Studies (GISS), in New York.

Using a GISS computer model that includes a variety of data gathered by NASA and other U.S. satellites, the researchers simulated realistic aerosol concentrations of important aerosol types in the atmosphere and studied the amount of light and heat they absorb and reflect over several regions around the globe.

Each area has a unique mix of natural and pollutant aerosol sources that produces different types of aerosols and causes complex climate effects. The industry and power sectors are particularly important in North America and Europe and produce large amounts of sulfur dioxide, while Asia has higher emissions from residential sources, which produce relatively more carbon-containing aerosols.

"Computer model simulations showed that black carbon in the Arctic, a potentially important driver in climate change, derives its largest portion from Southeast Asian residential sources," said Koch. "According to current model estimates, the residential sector appears to have a substantial potential to cause climate warming and, therefore, could potentially be targeted to counter the effects of global warming."

Black carbon absorbs sunlight, warming the atmosphere just as dark pavement absorbs more sunlight and becomes hotter than light pavement. It has a large influence on global climate, because winds transport approximately half of the black carbon aerosols produced in important aerosol source regions like Asia and South Africa to other parts of the world. When lofted above precipitating clouds, these aerosols can remain airborne for relatively longer periods. Some of these aerosols are carried to the Polar Regions, where they settle on the surface of ice or snow, absorb sunlight, and boost melting.

Most particles, especially sulfates produced from the sulfur dioxide emissions of factories and power plants, are light-colored and tend to cool the atmosphere by reflecting sunlight or making clouds more reflective. Computer

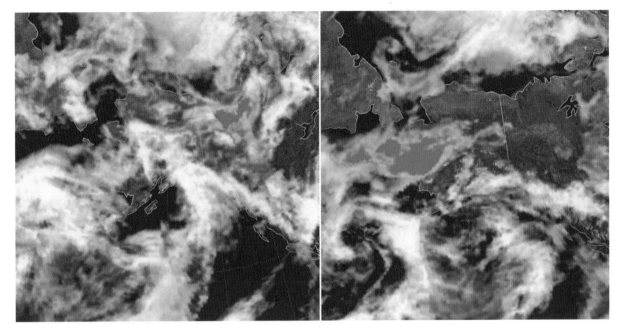

This pair of images from the Ozone Monitoring Instrument (OMI) on NASA's Aura satellite shows smoke measurements over Alaska and western Canada on August 15, 2005 (left) and August 21, 2005 (right). Increasing amounts of smoke are shown as an aerosol index with shades of blue (little or no smoke) to dull red (thick smoke).

model simulations suggest this effect is especially heightened over parts of the Northern Hemisphere, including the Central United States. The study found, however, that sulfur dioxide emissions in Southeast Asia and Europe have a smaller impact on climate because atmospheric conditions in those areas are not as efficient at turning the emissions into sulfate particles.

"This research is only the first step in considering the impacts of aerosols from different sectors on climate," said Koch. "Aerosols have other effects, like altering cloud characteristics that influence precipitation and climate."

NASA Satellites Unearth Antarctic 'Plumbing System' and Clues to Leaks

Imagine peering down from aboard an airplane flying at 35,000 feet and spotting changes in the thickness of a paperback book on a picnic blanket in New York City's Central Park. If you believe this impossible, NASA satellites are doing the equivalent of just that. From nearly 400 miles above the Earth, satellites have detected subtle rises and falls in the surface of fast-moving ice streams on the Antarctic ice sheet, a capability that also offers scientists an extraordinary view of interconnected waterways deep below that surface.

"NASA's satellite instruments are so sensitive we're able to measure from space changes in the ice sheet's surface elevation of a mere 3 feet," said Robert Bindschadler, chief scientist of the Hydrospheric and Biospheric Sciences Laboratory at Goddard and co-author of a related study published in the February 16, 2007, issue of Science.

With the aid of the satellites, Bindschadler and a team of scientists led by research geophysicist Helen Fricker, of the Scripps Institution of Oceanography, in La Jolla, California, revealed a new three-dimensional look at an extensive network of waterways beneath an active ice stream that acts like a natural plumbing system, as well as clues to how leaks in the system impact the world's largest ice sheet and sea level. They also documented, for the first time, changes in the height of the ice sheet's surface as proof that the lakes and channels nearly half a mile of solid ice below filled and emptied.

"This exciting discovery of large lakes exchanging water under the ice sheet's surface has radically altered our view of what's happening at the base of the ice sheet and how ice moves in that environment," said Bindschadler.

Fricker, Bindschadler, and others spotted intriguing discharges of water from the lakes into the ocean. Their research has also delivered new insights into how much

Because Antarctica holds about 90 percent of the world's ice and 70 percent of the world's reservoir of fresh water, leaks under the ice sheet influence sea level and ice melt worldwide.

water leaks from these waterways, how frequently, and how many connect to the ocean. Because Antarctica holds about 90 percent of the world's ice and 70 percent of the world's reservoir of fresh water, leaks in this system influence sea level and ice melt worldwide.

The research team combined images from an instrument aboard NASA's Terra and Aqua satellites and data from NASA's Ice, Cloud, and Land Elevation Satellite (ICESat) to unveil a first-ever view of changes in the elevation of the icy surface above a subglacial lake the size of Lake Ontario that took place over a 3-year period. Those changes suggest that the lake drained and that its water relocated elsewhere.

To the naked eye, the surface of the ice sheet is very cold and stable, but the base of any of its ice streams is warm, enabling water melted from the basal ice to flow, filling the system's "pipes" and lubricating flow of the overlying ice. These waterways act as a vehicle for water to move and change its influence on the ice movement, a factor that determines ice sheet growth or decay.

"There's an urgency to learning more about ice sheets when you note that sea level rises and falls in direct response to changes in that ice," said Fricker. "With this in mind, NASA's ICESat, Terra, Aqua, and other satellites are providing a vital public service."

NASA Data Helps Pinpoint Wildfire Threats

NASA data from Earth observation satellites is helping build the capability to determine when and where wildfires may occur by providing details on plant conditions.

While information from sophisticated satellites and instruments have recently allowed scientists to quickly determine the exact location of wildfires and to monitor their movement, this geoscience research offers a step toward predicting their development and could complement data from National Oceanic and Atmospheric Administration (NOAA) weather satellites used to help calculate fire potential across much of the United States.

By studying shrublands prone to wildfire in southern California, scientists found that NASA Earth observations accurately detected and mapped two key factors: plant moisture and fuel condition—or greenness—defined as the proportion of live to dead plant material. Moisture levels and fuel condition, combined with the weather, play a major role in the ignition, rate of spread, and intensity of wildfires.

"This represents an advance in our ability to predict wildfires using data from recently launched instruments," said lead author Dar Roberts, of the University of California, Santa Barbara. "We have come a long way in just the past 5 to 10 years and continue to gather much better data on the variables critical in wildfire development and spread."

To find out how well NASA satellites could detect these factors, researchers first sampled live fuel moisture, a critical measure for assessing fire danger, from several different plant species in sites across Los Angeles County. The ground-based data, collected by the Los Angeles County Fire Department over a 5-year period, were then compared to greenness and moisture measures from NASA's Moderate Resolution Imaging Spectroradiometer (MODIS) and Airborne Visible/Infrared Imaging Spec-trometer (AVIRIS) instruments. The space-based data were often closely linked to the field measurements, suggesting the instruments can be used to determine when conditions are favorable for wildfires.

"Improving the role of satellite data in wildfire prediction and monitoring through efforts like these is critical, since traditional field sampling is limited by high costs, and the number and frequency of sites you can sample," said Roberts. "This new data on the relative greenness of a landscape also allows us to see how conditions are changing compared to the past."

The satellite data worked best on landscapes where one plant type was dominant. The amount of vegetation cover in an area and its growth rate also influence the reliability of satellite data for wildfire prediction.

The MODIS on NASA's Aqua satellite captured this photo-like image and fire detections, which are marked with red dots. Some of the fire detections appear only as "hotspots," places where MODIS detected unusually high temperatures, while other fires are producing obvious smoke plumes.

The study also found that in areas where branches and dead foliage often help spread fires, changes in the proportion of green vegetation to other plants may also indicate locations of potential fires, especially after moisture values fall below a critical level. The proportion of greenness determines the manner in which plants absorb and scatter sunlight and plays a major role in moisture retention.

Although scientists have long recognized the importance of moisture conditions in wildfire development, this research suggests that other variables may be just as significant. "While live fuel moisture values are critical in the development of wildfires, it's clearly not the last word. Even if vegetation is extremely dry, there are a number of other factors that influence whether a fire will develop and how quickly it spreads, including the ratio of live to dead foliage, plant type, seasonal precipitation, and weather conditions," said Roberts.

As researchers continue to better understand wildfire development, they are also creating fire-spread computer models that use wind speed and direction forecasts to determine where fires will travel. And in the near future, scientists will likely be able to map fire severity to get an indication of the overall impact of a wildfire on the landscape and environment, including the amount of carbon dioxide released into the atmosphere. As the data record from recent satellites continues to grow, scientists will also be able to better track historical changes that might modify fire danger to provide better information for decision makers.

Two NASA Technologies Inducted into the Space Technology Hall of Fame

On April 12, 2007, two water treatment technologies developed at NASA were inducted into the Space Foundation's Space Technology Hall of Fame. Johnson Space Center received the honor for its development of the Microbial Check Valve used in water purification, and Kennedy Space Center was recognized for the development of Emulsified Zero-Valent Iron (EZVI) technology used to clean contaminated ground water. Both technologies were featured in *Spinoff* 2006, and the EZVI technology was also recognized as NASA's "Government Invention of the Year" and "Commercial Invention of the Year" in 2005.

Developed at Johnson to provide microbial control for drinking water systems for the space shuttle and the International Space Station, the Microbial Check Valve is now an integral component in water purification systems in rural areas and developing countries around the world. Johnson engineers joined the Water Security Corporation, of Sparks, Nevada, and Umpqua Research Company, of Myrtle Creek, Oregon, as inductees for developing the technology. Retired NASA employee Richard Sauer received an individual award for his work on the Microbial Check Valve while he was the manager of Shuttle Water Quality at Johnson.

The EZVI technology is a cost-effective technology used to clean ground water contaminated by dense

Johnson Space Center's Michelle Brekke at the Space Technology Hall of Fame awards dinner with William B. Tutt, chairman emeritus of the Space Foundation (left) and Astronaut John Herrington (right), currently director of the Center for Space Studies at the University of Colorado.

chemical compounds. Engineers at Kennedy developed the technology to address pollution from chlorinated solvents used to clean Apollo rocket components. The environmentally friendly EZVI uses iron particles in an oil and water base that neutralizes the toxic chemicals. This technology is now used at both government and private industry cleanup sites.

Dr. Jacqueline W. Quinn, environmental engineer, and Kathleen B. Brooks, materials scientist, received individual awards for their work at Kennedy on the EZVI. Weston Solutions, of West Chester, Pennsylvania; GeoSyntec, of Guelph, Ontario; and the University of Central Florida, Orlando were also inducted for developing the technology.

Michele Brekke, director of Johnson's Innovative Partnerships Program office, and Dr. David Bartine, director of Applied Technology at Kennedy, accepted the awards on behalf of NASA at the Space Technology Hall of Fame dinner, the closing event of the Space Foundation's 4-day National Space Symposium held in Colorado Springs, Colorado.

The National Space Symposium is the premier event for the Space Foundation, a national nonprofit organization founded in 1983 and headquartered in Colorado Springs. The organization is a leader in space awareness activities, trade association services, research and analysis for the global space industry, and educational enterprises. Since 1988, the Space Foundation's Space Technology Hall of Fame, managed in cooperation with NASA, has honored 54 technologies, as well as the innovating organizations and individuals who transformed space technology into commercial products that improve life here on Earth. ❖

Google™ and Google Earth™ are trademarks of Google Inc.

Innovative Partnerships Program

The Innovative Partnerships Program creates partnerships with industry, academia, and other sources to develop and transfer technology in support of national priorities and NASA's missions. The programs and activities resulting from the partnerships engage innovators and enterprises to fulfill NASA's mission needs and promote the potential of NASA technology.

Innovative Partnerships Program

The Innovative Partnerships Program (IPP) provides needed technology and capabilities to NASA's mission directorates, programs, and projects through investments and partnerships with industry, academia, government agencies, and national laboratories. IPP has offices at each of NASA's 10 field centers, and elements that include: Technology Infusion, which manages the Small Business Innovation Research (SBIR) and Small Business Technology Transfer (STTR) programs and the IPP Seed Fund; the Innovation Incubator, which includes the Centennial Challenges and new efforts with the emerging commercial space sector; and Partnership Development, which includes intellectual property management and technology transfer. In 2006, IPP facilitated many partnerships and agreements:

- Over 200 partnerships with the private sector, Federal and State governments, academia, and other entities

- Over 50 license agreements with private entities for use of NASA-developed technologies

- Reporting of more than 750 new NASA technologies

- More than 400 agreements for commercial application of NASA software

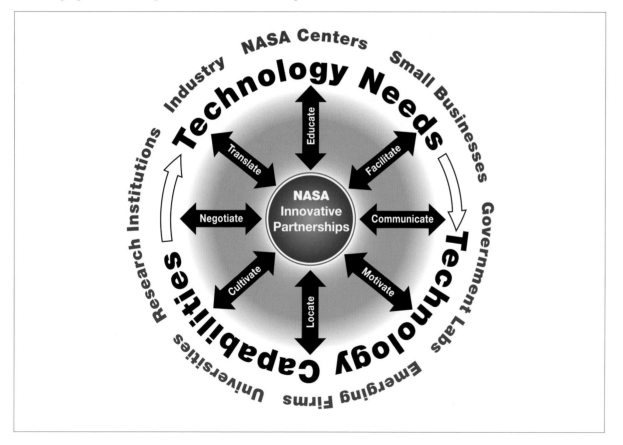

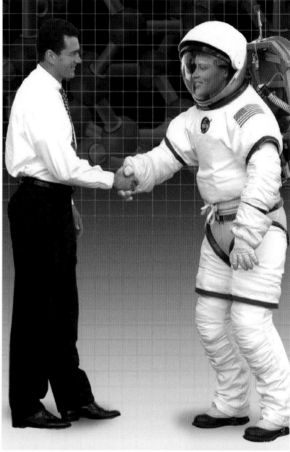

To complement the specialized centers and programs sponsored by the IPP, affiliated organizations and services have been formed to strengthen NASA's commitment to U.S. businesses and build upon NASA's experience in technology transfer.

The NASA **Small Business Innovation Research (SBIR)** program <http://www.sbir.nasa.gov> provides seed money to small U.S. businesses to develop innovative concepts that meet NASA mission requirements. Each year, NASA invites small businesses to offer proposals in

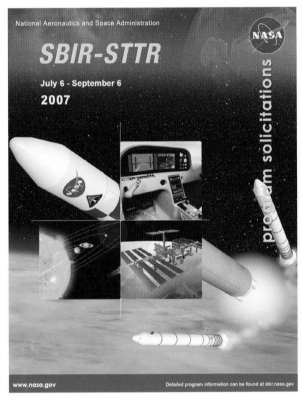

Small businesses develop technologies in response to specific NASA mission-driven needs, as presented in the 2007 NASA SBIR-STTR Program Solicitation.

response to technical topics listed in the annual SBIR-STTR Program Solicitation. The NASA field centers negotiate and award the contracts, and then monitor the work.

NASA's SBIR program is implemented in three phases:

- **Phase I** is the opportunity to establish the feasibility and technical merit of a proposed innovation. Selected competitively, NASA Phase I contracts last 6 months and must remain under specific monetary limits.

- **Phase II** is the major research and development effort which continues the most promising of the Phase I projects based on scientific and technical merit, results of Phase I, expected value to NASA, company capability, and commercial potential. Phase II places greater emphasis on the commercial value of the innovation. The contracts are usually in effect for a period of 24 months and again must not exceed specified monetary limits.

- **Phase III** is the process of completing the development of a product to make it commercially available. While the financial resources needed must be obtained from sources other than the funding set aside for the SBIR, NASA may fund Phase III activities for follow-on development or for production of an innovation for its own use.

The SBIR Management Office, located at the Goddard Space Flight Center, provides overall management and direction of the SBIR program.

The NASA **Small Business Technology Transfer (STTR)** program <**http://www.sbir.nasa.gov**> awards contracts to small businesses for cooperative research and development with a research institution through a uniform, three-phase process. The goal of Congress in establishing the STTR program was to transfer technology developed by universities and Federal laboratories to the marketplace through the entrepreneurship of a small business.

Although modeled after the SBIR program, STTR is a separate activity and is separately funded. The STTR program differs from the SBIR program in that the funding and technical scope is limited and participants must be teams of small businesses and research institutions that will conduct joint research.

SBIR/STTR Hallmarks of Success Videos are short videos about successful companies that have participated in the SBIR and STTR programs. Available online <**http://sbir.nasa.gov/SBIR/successvideo.html**>, many

The Lunar Lander Challenge, one of NASA's Centennial Challenges, is designed to accelerate technology developments supporting the commercial creation of a vehicle capable of ferrying cargo or humans back and forth between lunar orbit and the lunar surface.

of these videos also feature products that have been featured in the pages of *Spinoff*.

As another area of emphasis for Technology Infusion, the IPP established the **Seed Fund** with the following objectives:

- To support NASA Mission Directorate program/project technology needs

- To provide "bridge" funding to centers in support of Mission Directorate programs

- To promote partnerships and cost sharing with Mission Directorate programs and industry

- Leverage resources with greater return on investment

The Seed Fund issued a call for 1-year proposals that were scientifically/technically feasible, with relevance and value to NASA Mission Directorate programs and that leveraged the strengths and capabilities of an external partner.

In support of the IPP's Innovation Incubator element, **Centennial Challenges <http://centennialchallenges. nasa.gov/>** is NASA's program of prize contests to stimulate innovation and competition in solar system exploration and ongoing NASA mission areas. By making awards based on actual achievements, instead of proposals, Centennial Challenges seeks novel solutions to NASA's mission challenges from nontraditional sources of innovation in academia, industry, and the public.

To complement the IPP's Partnership Development efforts, the **National Technology Transfer Center (NTTC) <http://www.nttc.edu>** links U.S. industry with Federal laboratories and universities that have the technologies, the facilities, and the world-class researchers that industry needs to maximize product development opportunities. The NTTC has worked with NASA since 1989, providing the services and capabilities needed to meet the changing needs of NASA for managing intellectual property and creating technology partnerships.

On May 3, 2007, Peter Homer, of Southwest Harbor, Maine, won $200,000 from NASA for his entry in the Astronaut Glove Challenge, one of NASA's seven Centennial Challenges.

The **Federal Laboratory Consortium for Technology Transfer (FLC) <http://www.federallabs.org>** was organized in 1974 and formally chartered by the Federal Technology Transfer Act of 1986 to promote and strengthen technology transfer nationwide. More than 700 major Federal laboratories and centers, including NASA, are currently members. The mission of the FLC is twofold:

- To promote and facilitate the rapid movement of Federal laboratory research results and technologies from Federal research laboratories into the mainstream U.S. economy by fostering partnerships

and collaboration with the private sector, academia, economic development organizations, and other entities engaged in technology development.

- To use a coordinated program that meets the technology transfer support needs of FLC member laboratories, agencies, and their potential partners in the transfer process.

The road to technology commercialization begins with the basic and applied research results from the work of scientists, engineers, and other technical and management personnel. The NASA **Scientific and Technical Information (STI)** program **<http://www.sti.nasa.gov>**

provides the widest appropriate dissemination of NASA's research results. The STI program acquires, processes, archives, announces, and disseminates NASA's internal—as well as worldwide—STI.

The NASA STI program offers users Internet access to its database of over 3.9 million citations, as well as many in full text; online ordering of documents; and the NASA STI Help Desk (help@sti.nasa.gov) for assistance in accessing STI resources and information. Free registration with the program is available for qualified users through the NASA Center for AeroSpace Information.

The **NASA Technology Tracking System (NTTS)** **<http://technology.nasa.gov>** provides access to NASA's technology inventory and numerous examples of the successful transfer of NASA-sponsored technology. TechFinder, the main feature of the Internet site, allows users to search technologies and success stories, as well as submit requests for additional information.

The **Space Foundation <http://www.spacefoun dation.org>** is a national, nonprofit organization that vigorously advances civil, commercial, and national security space endeavors and inspires, enables, and propels tomorrow's explorers. The Space Foundation is a leader in space awareness activities, trade association services, research

and analysis for the global space industry, and educational enterprises that bring space into the classroom.

Working closely with NASA's Innovative Partnerships Program offices and public information offices, the Space Foundation each year recognizes individuals, organizations, and companies that develop innovative products based on space technology. These honorees are enshrined in the **Space Technology Hall of Fame <http://www. spacetechhalloffame.org>**. The ever-growing list of inductees showcases the significant contributions that space technology has made to improve the quality of life for everyone around the world.

For more than 3 decades, ***NASA Tech Briefs* <http:// www.nasatech.com>** has reported to industry on any new, commercially significant technologies developed in the course of NASA research and development efforts.

The monthly magazine features innovations from NASA, industry partners, and contractors that can be applied to develop new or improved products and solve engineering or manufacturing problems. Authored by the engineers or scientists who performed the original work, the briefs cover a variety of disciplines, including computer software, mechanics, and life sciences. Most briefs offer a free supplemental technical support package, which explains the technology in greater detail and provides contact points for questions or licensing discussions.

***Technology Innovation* <http://www.ipp.nasa. gov/innovation>** is published quarterly by the Innovative Partnerships Program. Regular features include current news and opportunities in technology transfer and commercialization, and innovative research and development.

NASA *Spinoff* <http://www.sti.nasa.gov/tto> is an annual print and online publication featuring successful commercial and industrial applications of NASA technology, current research and development efforts, and the latest developments from the NASA Innovative Partnerships Program. ❖

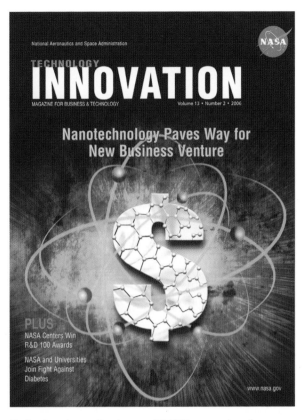

Technology Innovation is one of NASA's magazines for business and technology, published by the Innovative Partnerships Program.

NASA Innovative Partnerships Program Network Directory

ARC — Ames Research Center
CASI — NASA Center for AeroSpace Information
DFRC — Dryden Flight Research Center
FLC — Federal Laboratory Consortium
GRC — Glenn Research Center
GSFC — Goddard Space Flight Center
HQ — NASA Headquarters
JPL — Jet Propulsion Laboratory
JSC — Johnson Space Center
KSC — Kennedy Space Center
LaRC — Langley Research Center
MSFC — Marshall Space Flight Center
NTTC — National Technology Transfer Center
SF — Space Foundation
SPC/Fuentek — Systems Planning Corporation/Fuentek
SSC — Stennis Space Center

The 2007 NASA Innovative Partnerships Program (IPP) network extends from coast to coast. For specific information concerning technology partnering activities, contact the appropriate personnel at the facilities listed or visit the Web site: **<http://www.ipp.nasa.gov>**. General inquiries may be forwarded to the Spinoff Program Office at **<spinoff@sti.nasa.gov>**.

To publish a story about a product or service you have commercialized using NASA technology, assistance, or know-how, contact the NASA Center for AeroSpace Information, or visit: **<http://www.sti.nasa. gov/tto/contributor.html>**.

★ **NASA Headquarters** manages the Spinoff Program.

▲ **Innovative Partnerships Program Offices** at each of NASA's 10 field centers represent NASA's technology sources and manage center participation in technology transfer activities.

■ **Allied Organizations** support NASA's IPP objectives.

★ NASA Headquarters

National Aeronautics and Space Administration
300 E Street, SW
Washington, DC 20546
NASA *Spinoff* Publication Manager:
Janelle Turner
Phone: (202) 358-0704
E-mail: janelle.b.turner@nasa.gov

▲ Field Centers

Ames Research Center
National Aeronautics and Space Administration
Moffett Field, California 94035
Chief, Technology Partnerships Division:
Lisa Lockyer
Phone: (650) 604-1754
E-mail: lisa.l.lockyer@nasa.gov

Dryden Flight Research Center
National Aeronautics and Space Administration
4800 Lilly Drive, Building 4839
Edwards, California 93523-0273
Chief, Technology Transfer Partnerships Office:
Greg Poteat
Phone: (661) 276-3872
E-mail: gregory.a.poteat@nasa.gov

Glenn Research Center
National Aeronautics and Space Administration
21000 Brookpark Road
Cleveland, Ohio 44135
Director, Technology Transfer and Partnership Office:
Kathleen Needham
Phone: (216) 433-2802
E-mail: kathleen.k.needham@nasa.gov

Goddard Space Flight Center
National Aeronautics and Space Administration
Greenbelt, Maryland 20771
Chief, Office of Technology Transfer:
Nona K. Cheeks
Phone: (301) 286-5810
E-mail: nona.k.cheeks@nasa.gov

Jet Propulsion Laboratory
National Aeronautics and Space Administration
4800 Oak Grove Drive
Pasadena, California 91109
Manager, Commercial Program Office:
James K. Wolfenbarger
Phone: (818) 354-3821
E-mail: james.k.wolfenbarger@nasa.gov

Johnson Space Center
National Aeronautics and Space Administration
Houston, Texas 77058
Director, Technology Transfer Office:
Michele Brekke
Phone: (281) 483-4614
E-mail: michele.brekke@nasa.gov

Kennedy Space Center
National Aeronautics and Space Administration
Kennedy Space Center, Florida 32899
Acting Chief, Innovative Partnerships Program Office:
David R. Makufka
Phone: (321) 867-6227
E-mail: david.r.makufka@nasa.gov

Langley Research Center
National Aeronautics and Space Administration
Hampton, Virginia 23681-2199
Deputy Director, Advanced Planning & Partnerships Office:
Marty Waszak
Phone: (757) 864-4015
E-mail: m.r.waszak@nasa.gov

Marshall Space Flight Center
National Aeronautics and Space Administration
Marshall Space Flight Center, Alabama 35812
Acting Director, Technology Transfer Office:
James Dowdy
Phone: (256) 544-7604
E-mail: jim.dowdy@nasa.gov

Stennis Space Center
National Aeronautics and Space Administration
Stennis Space Center, Mississippi 39529
Manager, Innovative Partnerships Program Office:
Ramona Pelletier Travis
Phone: (228) 688-3832
E-mail: ramona.e.travis@ssc.nasa.gov

■ Allied Organizations

National Technology Transfer Center (NTTC)
Wheeling Jesuit University
Wheeling, West Virginia 26003
Darwin Molnar, Vice President
Phone: (800) 678-6882
E-mail: dmolnar@nttc.edu

Systems Planning Corporation
1000 Wilson Blvd
James M. Kudla, Vice President, Corporate Communications
Phone: (703) 351-8238
E-mail: jkudla@sysplan.com
Arlington, Virginia 22209

Fuentek, LLC
85 Goldfinch Lane
Apex, North Carolina 27523
Laura A. Schoppe, Director
Phone: (919) 303-5874
E-mail: laschoppe@fuentek.com

Space Foundation
310 S. 14th Street
Colorado Springs, Colorado 80904
Kevin Cook, Director, Space Technology Awareness
Phone: (719) 576-8000
E-mail: kevin@spacefoundation.org

Federal Laboratory Consortium
300 E Street, SW
Washington, DC 20546
John Emond, Collaboration Program Manager
Phone: (202) 358-1686
E-mail: john.l.emond@nasa.gov

NASA Center for AeroSpace Information
Spinoff Project Office
NASA Center for AeroSpace Information
7115 Standard Drive
Hanover, Maryland 21076-1320

Daniel Lockney, Editor/Writer
Phone: (301) 621-0224
E-mail: dlockney@sti.nasa.gov

Gareth Williams, Editor/Writer
Phone: (301) 621-0223
E-mail: gwilliams@sti.nasa.gov

John Jones, Graphic Designer

Deborah Drumheller, Publications Specialist